AF711821

PARK CITY

The old Silver King Coalition Mines building in the background has become the symbol of today's new Park City.

Diggings and Doings in

PARK CITY

Revised and Enlarged

Third Edition

Raye Carleson Ringholz

Portfolio: Harry Harpster

Photographs: K. C. Price

Introduction: Lowell Thomas

Western Epics

Salt Lake City, Utah

Copyright © 1972 by Raye Carleson Ringholz
Standard Book Number 0-87480-063-3
Library of Congress Catalogue Card Number 72-88556
Printed in the United States of America

No part of this work may be reproduced or utilized in any form or by any means, electronic or mechanical, including xerography, photocopying, microfilm, and recording, or by any information storage and retrieval system, without permission in writing from Western Epics, 254 South Main, Salt Lake City, Utah 84101.

Contents

Introduction
by Lowell Thomas

If you are keen about a place you want to keep coming back. That's what's been happening to thousands of people at Park City in recent years and that's what's been happening to me in Utah for over sixty years, since 1914.

In the thirties it was the skiing that lured me to the Wasatch Mountains — the feathery "powder" skiers go half-mad over. Since those late thirties I've rarely missed a season. Maybe I'm a sorta halfway citizen of Utah. At any rate, I feel at home here.

Of course there's lots more than skiing in "them thar hills." Where once there was silver galore, there is still the enduring legend of boom days, which makes the rebirth of rip-roaring Park City doubly exciting. I happen to be an ex-gold miner myself, having spent much of my youth in one of the fabled gold mining camps of Colorado where my father was a mining surgeon. From the old-timers I heard endless tales of mother lodes and the adventurous days when men worked with jack drills and dynamite in candlelit drifts and stopes far under the ground. Indeed, I was one of them.

Raye Carleson Ringholz brings this to life in her book, *Diggings and Doings in Park City* — doing it again in this third edition. The fascinating history is still there — discovery and boom days, underground ghosts and Main Street characters, the fun and frolic, the tragedy and heartbreak of the early West. But there's more.

We read of the bizarre events of Park City's first election; we take a closer look at the successes (and some scandals) of the

"Old Park City Bunch"; we meet Mother Urban, the 200-pound madam with a wooden leg; and we even get a glimpse of more recent prohibition days. We also have an updated account of the current "doings" which have been expanding dramatically — condominium villages, art festivals, theatre groups, restaurants and night life, and as lively a crowd of young people as you will encounter anywhere this side of Timbuctu and Kathmandu.

This third edition adds a new chapter — Diggings Now. The Ontario Mine, first permanent producer in the area, is at it again, — with different equipment, different methods — and even women working underground. K. C. Price captures it all with her camera, taking you to a stope 2,400-feet into the mountain where a young lady breaks rock with a 70-pound jackleg drill; letting you balance on a two-by-four over rushing water in the mine tunnels; giving you a first-hand look at modern ore crushing methods in the new Ontario mill on top.

Harry Harpster's Portfolio of photographs is here, too, reflecting the warped facades of historic buildings, details of delicate Victorian "gingerbread"; the faces, the action, the atmosphere of "now" combined with reminiscences of "then."

A collection of old-time photos completes the story and contributes an invaluable record of the days of yesteryear.

As an extra bonus, we also have popular "boom-town" recipes representing the mixed bag of nationalities in the old camp.

So let's be a part of it. Let's head for Park City.

Park City...Then

View of Park City around the turn of the century.

Mining in the days before hardboiled hats and acetylene lamps.

A stope in the west drift of the Anchor Mine.

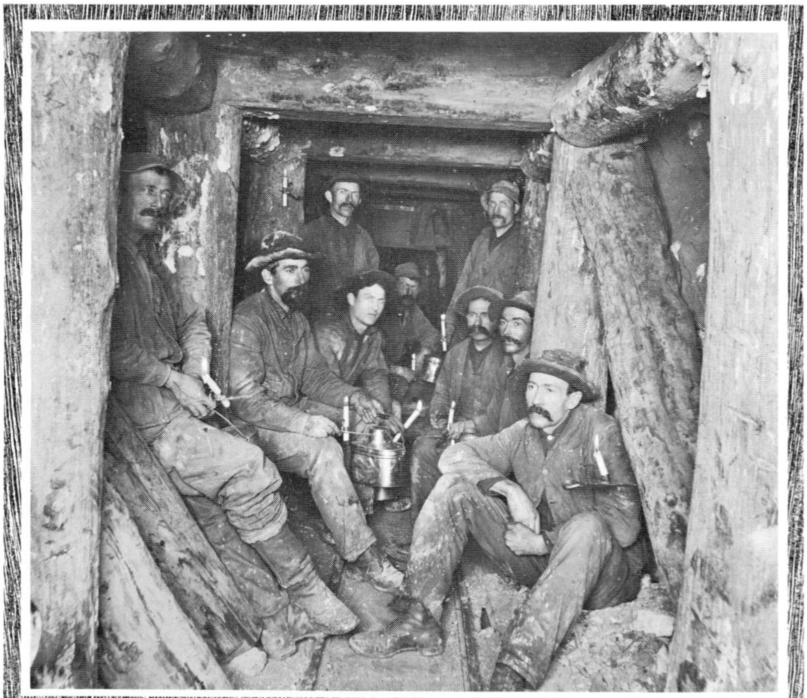

Diggings

The Wasatch Mountains held their secret many centuries. The sheer, barren crests shrouded their riches from Indians who came to refresh themselves with the water of Silver Creek and search among golden wildflowers for supplies of crickets and berries to feed their tribes in winter.

The rocky peaks gave no hint of an El Dorado to Spanish explorers seeking a route from Santa Fe to their new Pacific missions or to mountain men bent on setting traps for beaver rather than tapping the hills for treasure.

If the high country suggested fortune to Brigham Young, it was virtually ignored. Brigham delivered the Latter-day Saints from eastern religious persecution and instructed them to build a self-sufficient farming community beneath the western shadow of the Wasatch range in the neglected Great Salt Lake Valley. Although he dispatched official exploration parties to seek utilitarian coal, lead, and iron deposits, he forbade his followers to prospect for precious metals which might contaminate them with "gold fever" and would surely bring a rush of Gentiles to disrupt the exclusively Mormon settlement.

But Parley P. Pratt, another pioneer leader, remembered how welcome a serene mountain meadow had appeared on the last hitch of that long, arduous trek to the Promised Land. He rode up the canyon again to explore it and apprised Brigham Young of its fine grazing and timberland and its ideal location for controlling emigrant traffic into and out of the valley. Some time later, the land (known as Parley's Park) was granted to Samuel C. Snyder, Heber C. Kimball, and Jedediah M. Grant. Pratt built a toll road into the city, Kimball and Grant stretched another to the east, Snyder erected a sawmill, and the district spawned its first community, Snyderville, two miles below the present site of Park City.

Not until twenty years later was it hinted that the southern end of this high-lying mountain prairie, thirty miles east of Salt Lake City, Utah, would become one of the greatest bonanza camps of all time — Park City, a town whose silver mines would yield over $400,000,000, pay dividends of $90,000,000, and create some twenty-three millionaires.

Drilling with single jacks in the Daly-Judge Mine.

Water was an ever-present problem in the mines.

Surveying crews saved steps by using the miners' hand-pumped cars.

They call Colonel Patrick Edward Connor "The Father of Utah Mining." Stationed at Camp Douglas on the east bench of Salt Lake City, Connor's California Volunteers were ostensibly ordered to protect the Overland Mail route from Indian attack, but they were equally concerned with keeping an eye on the Mormons.

Ever since 1857, in fact, when President James Buchanan got word that "the civil laws of the Territory are over-shadowed and neutralized by a so-called ecclesiastical organization, as despotic, dangerous, and damnable as has ever been known to exist in any country . . . ," the United States Army had frowned upon the Land of Zion.

W. W. Drummond resigned his post as justice of Utah Territory after stating his conviction that "there is a secret oath-bound organization among all the male members of the Church to resist the laws of the country, and to acknowledge no law save the law of the 'Holy Priesthood,' which comes to the people through Brigham Young direct from God. . . ."

So President Buchanan named new officials and, to emphasize their authority, granted them military escort under Colonel Albert S. Johnston. But Johnston's mission came unannounced to the wary Mormons who considered it an "invasion." Unrest intensified and a small-scale Utah War resulted.

In 1860, Washington's attention was diverted by rumblings of Civil War, and Johnston (who had been promoted to general) was recalled. Two years later, Colonel Connor was assigned to establish the new Camp Douglas in case the Latter-day Saints decided to take advantage of the southern hostilities and declare themselves an independent commonwealth.

When the colonel arrived, he stood on the banks of the Jordan River west of town and vowed to cross over even "if Hell yawned below him." His fierceness was for naught. He was received in silence and, quite without incident, encamped at his hillside stronghold to train cannons upon the little settlement.

The volunteer soldiers soon looked to the high country. They hankered to dig into the mountains so similar to those of California, where regular army duty had kept them from panning gold. Connor, now a general, wasn't averse to a strike which would flood Utah with Gentiles and possibly solve his Mormon problems, so he issued permission for prospecting during off-duty hours. July 1864, he wrote the war department:

> As set forth in former communications, my policy in this Territory has been to invite hither a large Gentile and loyal population, sufficient by peaceful means and through the ballot box to overwhelm the Mormons by mere force of numbers and thus wrest from the church — disloyal and traitorous to the core — the absolute and tyrannical control of temporal and civil affairs, or at least a population numerous enough to put a check

The first Silver King Mill.

A surveying party.

Raw ore was placed on concentrating tables which vibrated to separate the heavier metal from light waste products.

on the Mormon authorities. . . . With this view, I have bent every energy and means of which I was possessed, both personal and official, towards the discovery and development of the mining resources of the Territory, using without stint the soldiers of my command whenever it could be done without detriment to public service.

Just a year after Connor's arrival, Utah's first mineral deposits were located in Bingham Canyon, on the eastern slopes of the Oquirrh range. Encouraged soldiers picked their way through other hills. There was a strike at Alta and exploration in Little Cottonwood Canyon. Then, one cold October afternoon in 1868, three soldiers, checking a ridge above Parley's Park, found an outcropping of yellow metal-bearing quartz. They chipped off a piece and grinned as it lay heavy in their hands. A snowstorm was threatening and it was turning dark, so one of them pulled a red bandana from his pocket and tied it to a bush to mark the place. They hiked ten miles down valley to catch Kimball's stagecoach and vowed to return next summer. The ore sample from the future Flagstaff Mine was later assayed at ninety-six ounces of silver per ton, with good portions of gold and lead.

Some vouch that this was the first strike in the Park City district; others maintain that discoveries made in 1869 by Rufus Walker and Ephraim Hanks deserve the distinction. Laxity of early prospectors in reporting to officials triggered such confusion.

Actual work on the Flagstaff didn't begin until two years later. The soldier-miners were hacking out a fifty-foot trench when they were approached by James M. Kennedy who offered them $5,000 on the spot. The deal was made and Kennedy filed the first legal record of a Park City claim.

Later Kennedy sold the mine to E. P. Ferry for $50,000, and within a short time the property, reorganized as the Marsac Silver Mining Company, had shipped $100,000 worth of high-grade ore from the diggings.

The news was out. Brush shanties, tents, and claim markers freckled the landscape as prospectors tapped the hills.

The Mormons still couldn't participate in mining precious metals, but they weren't forbidden to share the bonanza in other ways. Farmers from Kamas and Heber valleys sold dairy food and produce. Bill Kimball operated a livery stable and offered daily stage service to Salt Lake City. Sam Snyder's sawmill furnished timber for mine shores and shacks, and his brother George — with wife Rhoda — enlarged their home and took in boarders. "Aunt Rhoda's" reputation for good home cooking thrived and, on July 4, 1872, George hoisted a flag made of white sheeting, a red flannel baby blanket, and a blue silk handkerchief, to christen the burgeoning community "Park City, Utah."

View of old Ontario Mill which was later torn down.

Silver King Mine.

The Marsac Mill before it was destroyed in the fire.

David Keith, another miner turned millionaire.

Huge timbers were lowered into the mines to support tunnel walls.

The big strike came that same year. Reports conflict as to whether it was Rector Steen, a California Gold Rush veteran, or Herman Buden who accidentally stumbled onto a small knob sticking out of the ground in Ontario Canyon that led to a rich vein of silver ore, but both men and their partners, James Kane and Augustus Dawell, were elated when samples assayed 100 to 400 ounces of silver per ton.

At the same time San Francisco financier J. B. Haggin and George Hearst — father of publisher William Randolph Hearst — took time from dealings in Virginia City, Nevada's, Comstock Lode to cast interested eyes toward Park City. They sent a scout to check out the Hawkeye-McHenry mines in McHenry Canyon, which were rumored to have had an average value of $200 per ton for sixty-five samples from a ninety-foot tunnel.

Park City is considered a "wet district" because of the region's heavy precipitation which is sponged up by red shale and massive quartzite forming underground streams. When Hearst and Haggin's scout arrived at the Hawkeye-McHenry diggings, he found a heavy flow of water and spotty ore occurrences. He was about to reject the whole proposition when he happened over a ridge into Ontario Canyon, where Steen and his men were trenching their property. Dismounting from his horse, he took a sample for assay. Before thirty days had passed, Hearst-Haggin had purchased the Ontario for $27,000.

The initial year of operation was a frustrating one, however, because the rich vein was completely lost. After much futile searching, the mine was almost abandoned, but a determined worker named Patrick Kerwin begged more time and finally relocated the ore. With this beginning, the Ontario Silver Mining Company became the first permanent producing mine in the area, producing $50,000,000 worth of ore and dividends of $15,000,000 recorded to 1956.

It was this Park City money which was to lay the basis for the fabled Hearst fortune, but fate kept the millionaire from a bigger take. In the 1920s the Park Utah Mining Company was driving a crosscut from the Ontario No. 2 drain tunnel when the miners discovered an ore body which was to produce $15,000,000 and pay $5,000,000 in dividends. The vein extended to the 450-foot level of the old Hawkeye-McHenry mine; Hearst's scout had seen that mine down to the 400-foot level, missing the ore by 50 feet.

For a long time it was thought that the Ontario fissure was the only mine in camp and few had the pluck to look elsewhere. But J. J. Daly, a miner at the Ontario with a theory that the ore extended west, struck with the Daly Mine. E. P. Ferry, theorizing that if the Ontario ores went as far as Daly's property they would extend even farther, found the Anchor Mine. Then came the Daly-

West, Woodside, Mayflower, Silver King, Thaynes, Alliance, Crescent, and others.

Hundreds of lodes were clawed for rich silver-, gold-, lead-, and zinc-bearing ores. Thousands of dollars were poured into each venture before there were any results; the Daly Mine, for instance, invested $200,000 before a single pound of ore was uncovered. But those who struck hit big. The *Park Record* for October 24, 1896, boasted that the Ontario's record of dividends to date totaled $13,295,000, without a single assessment levied on its stock. The Daly Mine's earnings were $2,850,000 with a total output upward of $14,000,000.

The Crescent produced 1,000 tons of ore and paid two dividends; the Woodside had already resulted in half a dozen millionaires, one of whom was Thomas Kearns, an original founder of the fabled Silver King Mine, which ultimately paid its owners over $5,000,000.

But it wasn't all easy going. Mining equipment was primitive in the early days; metal instruments to pick, pry, chip, and pound; two men working together with a double-jack, one turning the drill while the other sledged; blowing out with dynamite.

Before the use of carbide lamps in 1912, each miner stabbed a dagger-like candleholder into the timber where he worked and used a daily supply of three candles for his light. Besides illumination, the candles served as warnings for threatened explosion. If the flame flickered, everyone dropped his tools and ran for the surface.

They called their work clothes "diggers," and when felt hats proved unsatisfactory under falling rocks, they changed to head coverings of yellow oilcloth with a hard, protective undercoating, which caught the name "hard-boiled hats."

It was a dangerous business. Several times snowslides buried miners alive. Two men were once crushed to death by a cage that plunged down the shaft when the hoist operator's attention was diverted momentarily. Another time a man amputated his arm. Many developed "miner's con" from breathing dust.

The worst mine disaster in Park City's history occurred July 15, 1902, when the powder magazine on the 1,200-foot level of the Daly-West exploded killing thirty-four men. Although the cause was never determined, it was suspected that the "powder monkey" on duty either dropped his candle or a cigarette into the magazine.

Horses, with candles tied in tin cans around their necks, pulled ore cars along tracks in the mine tunnels. Often there were thirty-five animals underground at a time. Only large three-year-olds of fine quality were purchased, and never those that had previously been hitched to a wagon because they had to be taught to keep ahead of the trains regardless of terrain so that the cars would not overtake them on downgrades.

The famous Cornish Pump.

Office of the Daly-West Mine.

Drilling underground was dangerous work.

The Ontario Mill Boarding House made miners comfortable close to their work.

Mine foremen were the only ones to enjoy the luxury of a tablecloth in the Daly Mine dining room.

Getting horses into the mine was quite a production. They were walked through dark tunnels to the shaft where they were bound in slings of rope and canvas — front legs to sides, hind legs to belly — then dangled lengthwise from the bottom of a hoist platform and lowered into the diggings. Because of the horses' cramped position, their breathing was restricted and there was danger that they would suffocate. As a precaution, they were denied food and water for three days prior to the trip, and the lowering process was accomplished within fifteen minutes.

Once in the mines, the horses lived quite comfortably in cave-like stables and remained underground as long as eight years. They became regular pets to the miners, who called them "hay-burners," and were even accustomed to sharing plugs of chewing tobacco. Cats were brought into the mines, also, to "mouse" the hay. They slept on the horses' backs at night for warmth. Contrary to popular belief, the horses did not turn blind. When time came to bring them to the surface, they were hoisted at night or blindfolded in daytime to allow their eyes to readjust to sunlight.

Hard-rock mining was rugged work and icy waters seeping underground often threatened to flood the mines. The Ontario pioneered a deep drainage system powered by steam boilers fired with cordwood cut from local forests. The owners ordered a $110,000 Cornish Pump for No. 3 shaft, and Parkites took pride in a machine unrivaled in any other western community. The flywheel weighed seventy tons and was thirty feet in diameter; the pump rod of Oregon pine was 1,060 feet long and sixteen inches square; water could be lifted from the 1,000-foot to the 600-foot level at the rate of 2,560 gallons per minute.

Legend has it that one day the pump stalled. The Ontario owners were desperate until they heard of a Scottish engineer who had worked with Cornish pumps before. A runner was sent and the Scot obliged by climbing the flywheel, pondering a minute, then hefting one mighty blow with a sledge-hammer. The wheel started.

The owners lavished praise on the Scotsman and asked his fee. When he demanded $1,000 for the job, the outraged management cried for an itemized bill. Without hesitation, the Scot replied, "For hitting the wheel with a hammer, 50 cents; for knowing where to hit the wheel, $999.50!"

Fabulous as the Cornish Pump had appeared, it was still insufficient. In 1888 work began on a 15,490-foot long Ontario Drain Tunnel to eliminate water at 1,500 feet. The timbered hole was 9 feet high, 4½ feet wide at the top spreading to 5½ feet at bottom, and contained a flume 5½ feet wide and 21 inches deep that was cut out of solid rock and capable of carrying 13,000 gallons of water per minute. The flume was covered with planking to serve as a tow path for underground animals.

Horses pulled ore carts through the mines.

Miners tucked three candles into their pockets, held their lunchbuckets, and posed for posterity.

The blacksmith shop was an important part of every mine.

Old-fashioned candles provided safety warnings as well as light.

Six years of construction were thwarted by fires, flooding, and swelling ground. After five years' digging in one direction, excavation started at the opposite end and, despite several miscalculations, the two points finally met head-on. The line was so straight that a person standing at the entrance could see daylight from the exit of the tunnel.

Foreman J. H. Keetley was ecstatic. He had the news telephoned to his wife, who immediately ran up the flag and instructed the engineer to blow all whistles. By the time Keetley appeared topside, a crowd had gathered and a "demijohn" was produced to fittingly celebrate the occasion. Everyone filed inside for a look, Mrs. Keetley having the distinction of being the first lady to do so. The drain tunnel's success made scrap of the Cornish Pump and led rival mines to instigate similar tunneling.

Miner in a stope of the Silver King Mine.

Nine miners at a time could descend in the Daly-West cage.

An atmosphere of stability was added to the community when the Marsac Silver Mining Company erected a $100,000 twenty-stamp mill in 1874. Previously, base ores had to be shipped to Liverpool, England, for reduction. That same year, the McHenry mill was built to treat chloride ores. Once milling facilities were initiated, other companies followed suit and built their own plants.

Raw ore was hauled fifty miles on animals' backs or oxcarts to the nearest smelter in Sandy, Utah, before opening of the Park City Smelting Company in 1880. Later plans to expand the plant never materialized because the ore declined in richness and rendered the smelting process unsatisfactory.

Plummeting silver prices in the nineties triggered the need for more economical methods of mining and treating low-grade ores. The Peck Concentrator added a new process of separating ore from waste rock by centrifugal force. The Concentrator worked with such thoroughness that values which previously escaped from mills and passed down Silver Creek (dubbed "Poison Creek" due to the pollution) were saved.

The Silver King caused quite a sensation in 1899 when it constructed an aerial tramway which transported ore downhill and coal uphill in a series of buckets depending from steel cables suspended on towers. Buckets were filled with ore at the mine, then carried down to railroad cars for shipment. It took only one man to work the automatic controls, and the tramway handled twenty-one tons of material per hour.

Most miners lived in company boardinghouses. Each man was issued a bed and locker in the Change Room. He brought his own bedding and was urged to report the presence of bedbugs or other vermin. The Silver King's Bunk House Rules cautioned, "Sociability and good fellowship is encouraged, but loud and boisterous

Large cables unwound on the hoisting drum.

Muckers removed waste material from mines.

talk, vulgarity or profanity, will not be permitted in the dining room."

The Park Utah Company had three boardinghouses for 275 employees. Company butchers cut meat and made sausage, and a first-class baking plant kept up a supply of breads and pastries. Thirteen Chinese cooks accommodated changing shifts with family-style meals four times a day, philosophizing, "Give them lots of thick soup and then, men, they don't eat so much!" Miners filled their own lunch buckets every day from a long table piled with sandwiches, boiled eggs, desserts, etc. When work was over, they returned their numbered lunch pails to be washed and laid out for refilling next shift.

One time it was suspected that the enterprising Oriental cooks at the Daly-West were selling rice saki to the townsfolk. The sheriff complained, so the superintendent did a bit of checking. When he asked his purchasing agent if he'd been buying much extra rice, the answer was, "Enough to feed the Chinese army!" The ration was cut and bootlegging operations ceased.

Board and room were $1.00 a day. Muckers were paid $2.75, miners $3.00, and timbermen $3.50 per shift. Payday was once a month.

The miners, primarily immigrant laborers representing many nationalities, wound through over 1,000 miles of tunnels and worked ten-hour shifts with only two days of vacation a year: Christmas and the Fourth of July. Unlike the mines of Virginia City, Nevada, where underground water steamed to 170 degrees, making fifteen-minute shifts barely endurable, the icy waters beneath Park City kept muckers hopping just to keep warm.

They were a superstitious lot. Just about everything was bad luck, from letting women in a mine to whistling on the job, dropping tools, or hearing a howling dog. If his candle snuffed out, a miner knew there was trouble at home; if it doused three times, someone was molesting his wife.

It was taboo for miners to give notice that they would quit the job on a certain day. Men who had confided their plans had been killed or injured while working their last shift. This happened so many times that those intending to leave cheated fate by quitting a day earlier than they had announced.

The mines were rife with mountain ghosts. Park City had a wicked dwarf whose arms were so long that he could buckle his sandals without stooping down, and so strong that he could pull himself up a ladder without using his legs. Many swore that this dwarf kicked out the rungs of ladders to block the escape of miners who were trapped.

There were "Tommy Knockers," other vindictive little creatures who made strange sounds of voices in the walls, of ghostly single-drilling or of tapping which prophesied bad luck to an individual or group of miners.

An Irishman named Quinn claimed to have seen a beautiful woman riding a white horse on the 200-foot level of Zev shaft in the old Alliance mine. She had long blond hair and radiant white flesh. Quinn had seen her several times, but she disappeared every time he tried to question her. Some said the woman was seeking her husband who had been crushed in an accident years before. (Later historians might wonder if the haunted man could have been the same Quinn mentioned in the *Park Record* November 28, 1896, after he had "... filled up on 'red eye' and settled himself on the Union Pacific track for a comfortable snooze.")

Park City was touted as the greatest silver camp in the world; everyone boasted that there was enough ore to keep things busy another hundred years. Economics and the human element dictated otherwise.

First came the Silver Issue. Enormous discoveries in the West during the 1860s and 1870s resulted in silver production catching up with demand, thus lowering prices. Furthermore, when England and Germany deserted the bimetallic standard for gold, the silver market was flooded. Congress ordered changes in the United States coinage system and substituted greenbacks for silver dollars.

A few years later, specie payment was resumed, but silver prices continued to decline. Western mining interests suspected a grand plot by eastern bankers to horde gold and depress the value of silver dollars, thereby playing into the hands of England and Germany whose insistence upon an exclusive gold standard relegated silver to fluctuations in price comparable to cotton, wheat, or other articles of commerce.

Westerners waxed optimistic in 1890 when the Sherman Silver Purchasing Act required the treasury to make monthly purchases that would virtually consume the entire output of their mines. But in 1893, Congress repealed the act. By mid-January, 250 Parkites were out of work. In June, silver fell to seventy cents an ounce and depression smothered the little camp. Many mines closed down. Only a few weathered the crisis by cutting wages and bypassing dividends.

The economy was braided with good and bad for several years. They were optimistic in 1896, then flattened the following year. A heyday hit at the turn of the century, the Park City Consolidated Mining Company blossomed in the late twenties, the Park Utah Mine at Keetley was developed after World War I, and the New Park, the area's major gold shipper, peaked in 1939. Then the depression of the early thirties and World War II combined with

Cooks preparing dinner for miners at a company boarding house.

The Anchor Mine concentrator from an 1891 annual company report.

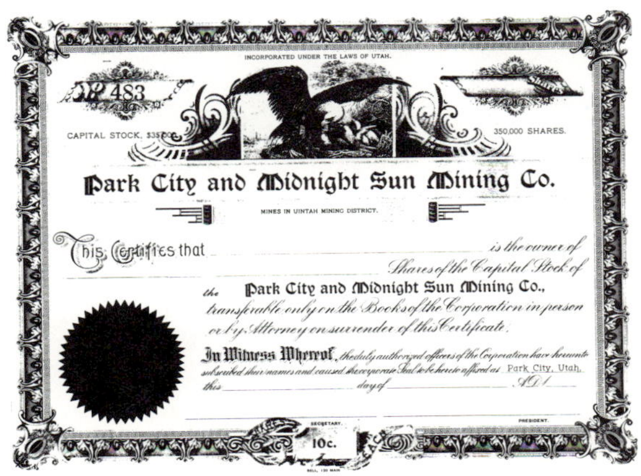

INCORPORATED UNDER THE LAWS OF UTAH.

No. 483

CAPITAL STOCK, $350,000. 350,000 SHARES.

Park City and Midnight Sun Mining Co.

MINES IN UINTAH MINING DISTRICT.

This Certifies that ______ is the owner of ______ Shares of the Capital Stock of the Park City and Midnight Sun Mining Co., transferable only on the Books of the Corporation in person or by Attorney on surrender of this Certificate.

In Witness Whereof, the duly authorized officers of the Corporation have hereunto subscribed their names and caused the corporate Seal to be hereto affixed at Park City, Utah, this ______ day of ______ A.D. 1

SECRETARY. PRESIDENT.

10c.

prohibitive pumping costs to force most of the mines to close.

In April, 1965, underground activity turned more scientific than commercial. A research team from the University of Utah (financed by an $850,000 grant from the National Science Foundation) started construction of a neutrino trap 2,000 feet below Treasure Mountain's summit, at the end of the three-mile Spiro Tunnel. The 2,000-ton detector probed mysteries of outer space by capturing tiny subatomic particles, called neutrinos, which travel throughout the entire earth at the speed of light. Through their experiments, researchers found neutrinos and muons (by-products of cosmic rays) and uncovered evidence suggesting the existence of another elusive and mysterious subatomic particle which might someday give us a better understanding of the forces of the universe.

The neutrino trap was terminated the summer of 1974, but the University started a new experiment that year at the 500-foot level of the Mayflower Mine. This research team is analyzing cosmic rays to determine if their intensity is the same from all parts of the sky or if there are any deviations in the form of a "cosmic ray wind."

But, it's not over yet. Unlike Virginia City — whose Comstock Lode produced a total of $320,000,000, paid $147,000,000 in dividends and $92,000,000 in assessments in less than twenty-five years, then petered out and left the city to dry up into a ghost town — Park City's yield of over $400,000,000 since 1872 dwindled but never stopped. Now, mining has resumed and a burgeoning resort has been developed. Prosperity in the Park is growing and flourishing in a variety of ways.

Heavy slickers worn by miners at the Ontario kept them dry while loading mule-drawn ore carts.

Marsac Mill retort room.

In 1894, mine housing parties were held uild homes for Park City's booming population.

Cosmic ray project in the Mayflower Mine.

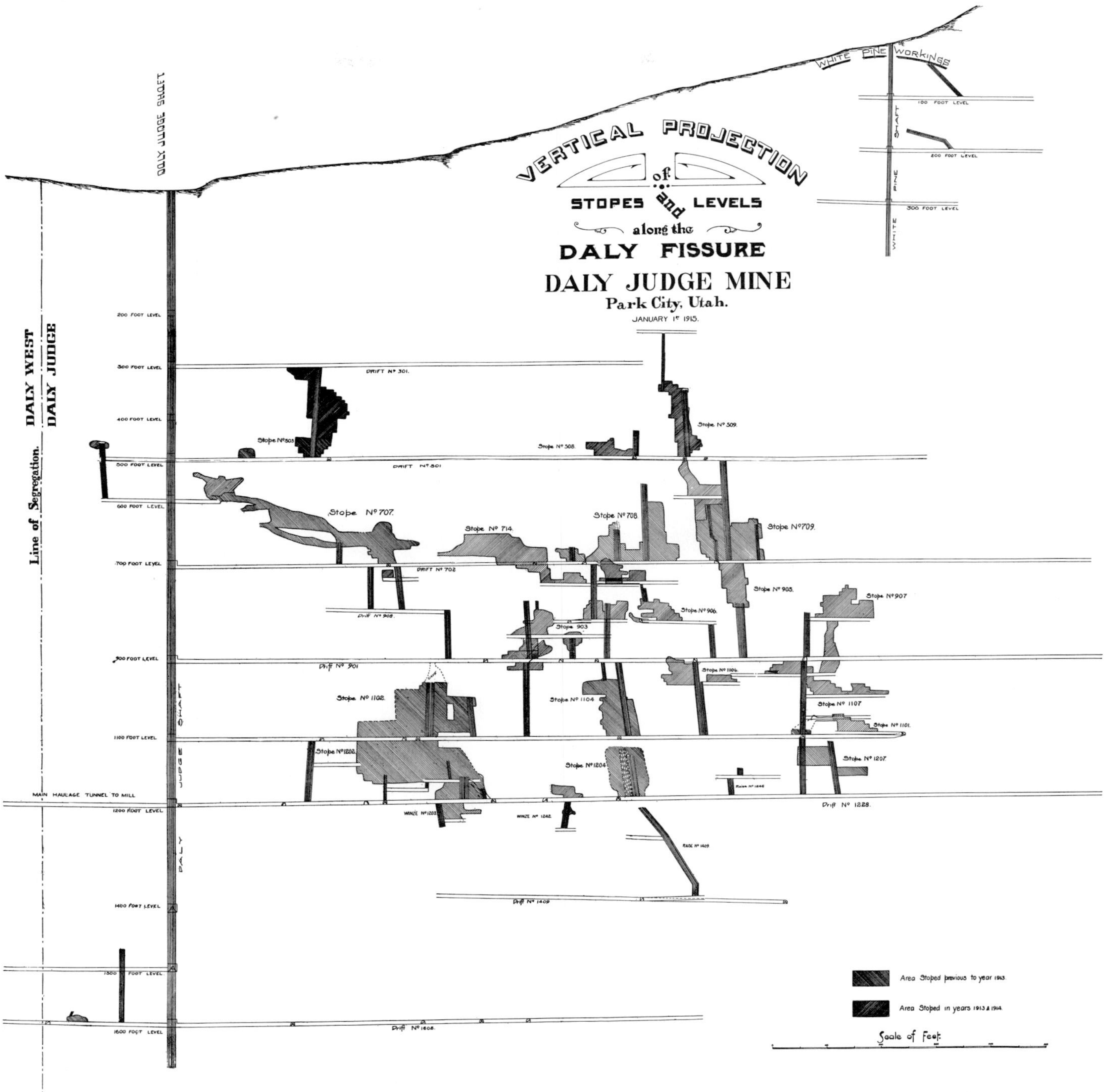

A vertical projection of the various stopes and levels of the Daly-Judge Mine.

Drinkers pause for the photographer in Sam De Angelas' Saloon.

Doings

They came on horseback, on blistered feet, or with donkey pack-trains following the canyon's natural streams. They contrived hasty shelters of brush and pine scraps then set picks ringing against outcrops of quartzite that might promise another Flagstaff, Ontario, or even better.

They came via Kimball's stagecoach or in wagons laden with supplies the prospectors needed but hadn't taken time from their diggings to buy. They wedged emporiums, blacksmith shops, livery stables, restaurants, and saloons into the narrow gap that slashed the Wasatch range and they converted solitude to commerce.

All kinds of people came to Park City: long-bearded prospectors and dream-filled youths with sights set on finding a fortune; hard-working family men seeking the uncrowded West to make a decent living; professional gamblers, anteing as long as the games paid off; elite Michiganers with classy houses on Park or Woodside Avenues; miners from the small town of Rossie, New York, making a bit of home across the way on Rossie Hill; "hot water boys" of Virginia City's played-out diggings, looking for the new bonanza; and characters such as "John the Baptist," a skeletal miner who preached hellfire and damnation or "Paddy the Pig," notorious for consuming a whole roast of beef at one sitting.

There were many nationalities. Paragons of mining skills were the Cornishmen, called "Cousin Jacks," who were sought for tough jobs, such as sinking shafts or driving tunnels through dangerous ground. During the first part of the nineteenth century, Cornwall produced most of the world's supply of copper, but discoveries in Spain and North America around 1860 flooded the copper market with cheaper metals and the Cornish mines folded. The skilled miners were forced to other parts of England, the Colonies, South America, and, inevitably, the Rocky Mountains, where they introduced many new techniques to the American mining industry.

They were a clannish bunch, distinctive because of their unique dialect and many colorful expressions. Their vocabulary included terms used in Freiburg, Germany, more than three centuries earlier — an offshoot from the influence of German miners brought to

Everything from Indian blankets to high-buttoned shoes was available at Welsh, Driscoll and Buck's bargain prices.

The dungeon under City Hall. Photograph by Harry Harpster.

Advertisement from the Park Record.

England by Queen Elizabeth I to teach Cornishmen to work refractory ores.

Cousin Jack lingo sparkled with imaginative terms. Oatmeal mush diluted with milk was "stirabout," thin meat turnovers were "Cornish pasties," and, when a miner's wife tucked a pasty into his lunch bucket, it was "a letter from home." It was they who coined "hard-boiled hats," "Tommy Knockers," "Mountain Ghosts," and when a mass of unproductive ground was in the midst of a mineral lode, it was "a horse in the lode." "Black Jack rides a good horse" meant that zinc ore gives a good promise of copper.

The nickname Cousin Jack probably derived from their habit of securing job replacements from their own family ranks or countrymen.

There was a Scotch contingent in Park City, too. One of the first Scots to arrive was John L. Street, who set his roots in a little glen near a lake by the McHenry mine. Others came to join in an aloof community that furnished the area with cabinetmakers, blacksmiths, and engineers.

More than half a million Irishmen left their homeland after the potato famine of 1847–54 and many of them found their way to Park City mines. Working as laborers or opening businesses and saloons, the Irish accounted for much of the camp's atmosphere and many of its fortunes. One of their favorite sons was Thomas Kearns, who first entered town with a pack on his back and left as a millionaire and a United States senator.

While the Irish are credited with much of the town's conviviality, the fact that the Catholic Church was first to build in town spoke well for their contribution to the stability of the community.

Park City's mountains were heavily forested in the early days. The lure of the West brought many Scandinavian lumbermen to fell trees for buildings and mine props. The northerners were industrious in mining company sawmills which dotted the back country.

When the Union Pacific and Utah Eastern Railroads completed their spur lines to Park City in December 1880, many Chinese laborers found themselves without work. Sensing a promising camp, the Orientals stayed on to open laundries, cook in boardinghouses, serve as waiters and porters, and sell fresh produce from their little gardens. Soon there was a Chinatown of fifteen or twenty shacks on the bank of Silver Creek. Old-timers remember slate smoke from "punks" and joss sticks and men with queued hair and silver coins tucked in their ears.

A bright red bridge used to span Chinatown and afford access to the fancier homes on Rossie Hill where Oriental peddlars delivered laundry or sold fresh fruits and vegetables. China Bridge was

Hodgson's Jewelry Store was a Main Street landmark for many years.

The Park City Furniture Store stacked its wares on the floor and hung them from the ceiling.

Mrs. Joffsin awaits customers in the old Park City Drug Store.

McClean's Grocery Store carried a little bit of everything.

The ornate Dewey Theatre was a great source of pride to Parkites.

Just hours after a capacity crowd left the Dewey Theatre, the elegant showhouse buckled under wet, heavy snow.

weakened by the fire of 1898 and was later replaced by a second, unpainted bridge, razed in 1954.

Chinese New Year was a city-wide celebration. Parents would take their children to see gaudy decorations and receive treats of nuts, candies, and fireworks.

Chinese funerals were also occasions for children in the camp. Hired bandwagons would shrill with Oriental music and lead processions of mourners to the cemetery where local kiddies would watch from behind fences as gifts of chickens, candies, and cakes were laid over the grave for the departing spirit. When the funeral was over, the children would feast and the Chinese could return the next day and be satisfied that the ghost had been appeased.

"The town of Park City is rapidly approaching the size and appearance of a healthy, compact, and permanent place," reported the *Salt Lake Tribune*, October 25, 1879. "Town lots are selling readily, and commanding a very flattering figure. Lots that two years ago sold for $10 to $20 are now bringing from $100 up."

By year's end there were 350 buildings for a population of 3,500 and, in 1880, the city sported boardwalks and the *Park Record*, the oldest weekly newspaper of continuous publication in Utah.

Certainly such a community should be incorporated, Parkites decided. They petitioned the 1880 Legislature but their bill succumbed to Governor George M. Emery's refusal to sign it because it was "inexpedient to incorporate small places."

They tried again in 1882. No one could deem their town a "small place," now, Parkites insisted. New-fangled gadgets called telephones had been installed and the Utah Eastern and Union Pacific railroad tracks rolled right up to Main Street.

This time the bill for incorporation passed. Judge Perks of Coalville proclaimed:

> In accordance with the wish of the people this Court has set the 5th day of August next, as the day upon which the first election, under the Park City charter, will be held and the warehouse of the Park City Smelting Co. has been designated as the place of holding said election. The following are to be elected:
>
> 1. Mayor
> 2. Alderman
> 3. Justice of the Peace
> 4. Constable
>
> The election will commence one hour after sun rise and continue until sunset....

The law provided that, in absence of a justice of the peace to name election judges, the first six electors to assemble on the appointed day would select them. However, when voters arrived at the warehouse and picked their judges, C. H. Wilhey declared

The Utah Central Railroad also provided contact with the Salt Lake Valley.

Doings at the Union Pacific Railroad Station were always an attraction for Parkites.

the building his private office and the ballot box his property. He refused the electorate admittance.

Undaunted, the new judges procured another ballot box and announced the vote would be taken at John L. Street's office.

Later, however, another party of voters arrived at the warehouse. This time Wilhey accommodated the crowd with ballot box and office. Brass bands blared and wagons carried voters to both polling places throughout the day. The "up-town party" at Street's claimed their election was official because they had opened the polls according to law and had been forced to change the location. The "warehouse party" maintained that they had operated in the designated location and used the official ballot box, so their count was the legal one. The *Park Record* commented on the problem: "Two voting places in one voting precinct is extremely rare and one of them is illegal. Which one?"

Both were illegal! In fact, NO election in Utah Territory during 1882 was legal. In March of that year, the United States Congress passed the Edmunds Law which provided for a five-man commission to establish election rules, appoint registration agents, and name judges. Until such matters were attended to by the committee, no election could be held and, as the five commissioners arrived in Salt Lake City on August sixteenth, Park City's election (or any other prior to that date) was void. Therefore, the impatient Parkites were unable to obtain their city charter until February 1884.

That was also the year they built their jail in City Hall. At one time the *Park Record* had condemned vigilance committees as a "curse to the community in which they existed"; in 1883, an editorial deplored the fact that "In two years, seven men have been sent to their graves by the hands of their fellow men. And the only punishment meted out to any of them is a short term in the penitentiary, where most of them fared better than they often would have fared."

Park City's first notable shooting was in 1877 when Bill Bennett and Jim Brasher took aim at each other. Bennett was discharged on self-defense and Brasher released on a $100 bond to await grand jury action.

In 1879, Louis Jungrau walked into Mrs. Power's boardinghouse and John Murphy growled that he could clear the room of "such men as had just entered" in a matter of minutes. He did. Jungrau recuperated in a Salt Lake City hospital and Murphy negotiated a successful escape while accompanying the constable to several saloons and a theatrical performance.

One of Utah's three lynchings happened in Park City. It seems that Matt Brennan and Will Gorlinski were helping E. M. Wheeler survey his claims when a rifle shot zinged from the direction in which "Black Jack" Murphy had disappeared moments before.

Brennan screamed, "My God, I am shot!" and fell dead, crying, "Murphy killed me!"

Evidence was strong against Murphy. Only a few days before, he had expressed "hard feelings" for Brennan and hinted that he'd like to "put him out of the way" and "had it fixed for a certain nameless party." He had borrowed a rifle and eight cartridges to go deer hunting, but, after the shooting, he locked himself inside of his cabin with the gun. Later, when he arrived in town with the weapon, there were only seven bullets.

Murphy was convicted on circumstantial evidence and sentenced to prison. The local populace was so incensed, however, that the sheriff decided to take him to the Coalville jail for protection. But a mob reared up to the cause of justice. They kidnapped a railroad engineer at gunpoint and forced him to run his locomotive to Coalville, then crept up on the sleeping jailer, snatched Murphy, and sped him to a rope at the top of a Park City telegraph pole. The jailhouse and its dungeon with leg irons (Utah's oldest Territorial jail still in use) was built the following year (1884).

Once-a-month paydays in the 1890s clattered to the accompaniment of popping corks and barroom pianos as bachelor miners, bent on ridding their pockets of unaccustomed jingling, swarmed into town from company boardinghouses.

Flip Wing, a Chinese gardner, shuffled across China Bridge to tally his month's bill that he had scratched in Oriental characters on the sides of customer's houses.

Jim Ivers ambled out of his livery stable and headed for the saloons to collect for horse rentals. He swung a neck yoke over his shoulder in case any debtors appeared reluctant with their cash. Jim really didn't expect to meet with trouble because his horses were very popular; they knew the boardinghouse trails and would wait for a drunken rider to remount when he slipped from the saddle en route home. On reaching his destination, all the bleary-eyed customer had to do was wrap the reins around the horse's neck and give him a swat to head him back toward the stable.

Bob Brogelmann kept his Palace Restaurant open day and night to dish out five-course meals including everything from German noodle soup to boiled ox heart with currant jelly, served "first class" at twenty-five cents.

Department stores plucked their share of the payroll, too. The P. McPherson Company took twenty percent off shoes for their Racket Sale, Julius Frankel's men's overcoats were featured bargains at $16 and up, while M. S. Ashheim Company touted suits for $10 to $20.

It was said, "If a fellow wants to become intoxicated he can start in at one end of the street, take a drink at each saloon or store that handles liquor, and before he is half way up or down, unless he has

Left: Waiters pour the bubbly in anticipation of the victory party in honor of Thomas Kearns' election to the United States Senate. Right: The celebration.

Tom Rease and Less Taylor mixed a little chatter with their barbering.

the capacity of a vinegar barrel, be gloriously intoxicated."

The Finns went to Finn Hall, Swedes to the Swedish Saloon, and P. J. Donovan advertised: "A peacock has a fine tail, but it don't compare with the cock-tails! prepared at P. J. Donovan's." They should have been good; P. J. had plenty of competition! Of Main Street's twenty-seven saloons, names like Morrison's, Riley and Towey's, Pape and Bowman's, Hallinan's, Fisher and Leahy's, Coffey and Farrell's, Kelly's, Fitzpatrick and McGowan's were predominant.

Blackjack tables attracted all-night wagering in back rooms. Brass-rail betting out front livened to talk of the race between "Spider Horse" and "Indian Dan" on a track near the Ontario drain tunnel, or to Ed Kelly and Will Chamberlain who challenged Duncan Gillis and Bill Rosevear $100 for a double-hand drilling match on a block of granite.

Livelier entertainment was found in the row of houses in Deer Valley; red lights winked and gay-ladies offered respite from a hard month's labors in the mines.

Sometimes it wasn't so restful, however; such as the night Louis Paradise stuck the muzzle of a 41-calibre Colt revolver in the Madam's breast when she tried to have him bounced. The gun misfired and the Madam got away, but the newspaper observed that "Louis is in serious trouble."

Threats came from both directions on "the row." One young man tried an unannounced leave-taking, but his "giddy girl" flashed a gun protesting — no money, no goodbyes. He offered her his watch, but when she noticed he had a second, she demanded both timepieces. The boy complained to the police but got little sympathy.

Fred Flint was a gambler of patrician Boston background who was widely respected because he never cheated at faro. When he first came to town, he had a girl over on the line, but in time he tired of her. Gentleman that he was, he told the lass. She took the blow quite bravely and suggested one last friendly drink. Fred obliged, but was appalled when the girl collapsed and died before they could rush her to the hospital. She had flavored his cocktail with arsenic and the glasses got mixed.

The police were kept busy "pulling the row," raiding hootchie-kootchie shows, investigating holdups, robberies, and "general cussedness," and there were times the leg irons in the dungeon below city hall were used. But for the most part, rowdies were short-time drifters and Park City was a settled community.

There was a police department, a fire department, churches and schools, telephones, railroads, and a thriving business and professional community.

St. Mary's of the Assumption Catholic Church was built in 1883

The Second China Bridge (or Fourth Street Bridge) over Poison Creek.

Picnics in the surrounding canyons were popular with Parkites.

Smith, Brian and Goodwin's grocery was a popular meeting place at the turn of the century.

Every June thirteenth, Parkites celebrated the annual Miners' Union Parade down Main Street.

and is the oldest Roman Catholic Church building in the state.

Protestant churches of many denominations were also active from early times, but because of Brigham Young's official attitude against mining, there were few Mormons in the mountain town. Consequently, Park City became a Gentile island with Catholics and Masons binding in a Loyalty League which was intended to eradicate Brigham's doctrines by peaceful means but with lawful force.

The few Mormons doing business there kept their beliefs to themselves, as evidenced in the *Park Record*, January 29, 1887. "Last fall a branch of the Mormon church was established in Park City with Gad Davis and P. W. Timms at the head of the rickety craft. Meetings were held in Erickson's cellar on Park Ave. and in the rear of Hop Chong's 'washee.' But it no longer exists in Park City."

Townsfolk found entertainment in many cultural pursuits. With subscriptions amounting to $30,000, they built a brick opera house, lavish with arches and filigreed stonework. The Payton Company and A. Y. Pearson's traveling group came to town with "The Great Four Act Western Mining Melodrama, 'The Golden Giant,'" or "Under the Gaslight." Edouard Remeny, the famous violinist, was booked as well as the Salt Lake German Concert Company or Mademoiselle Lira in "her grand transformation Serpentine Dance."

Professor H. B. Younger's Grand Balls were always great occasions. The good professor was ever ready with instruction in the latest two-step quadrille, and he staged special matinee balls for the youthful set.

Summer was a time for family outings and a day at the Great Salt Lake was a special treat. They would board the Denver and Rio Grande train at Park City, chug down the old Parley's Canyon track to Salt Lake City, then catch the open-air train to Saltair. After a day of band concerts, picnics, games, and floating in the brine, they would rattle back up the mountain.

Or families would take wagons to the Hot Pots resort in Heber Valley. There was warm mineral water bathing, picnic grounds, and everyone would climb the limestone craters to peer into vapors rising from volcano-like holes.

Old time Parkites still remember the excitement when John L. Sullivan came to town and the time William Jennings Bryan campaigned on the Silver Issue from the top of a ladder in front of the Union Pacific depot.

There were parades every Fourth of July. The Park City Independent Band set the pace and youngsters were treated to everything from pie eating contests to shinnying a greased pole, with fireworks and prizes galore.

Fire erupted at the head of Main Street on September 13, 1890. The business section suffered a $10,000 loss and was nearly destroyed. Editorials urged the City Council "to make some sensible moves toward protecting the town from total destruction by fire. It is high time that the Chief should get out the hose companies for drill," they continued, "so that the firemen will become familiar with the apparatus and know how to use it; know where to find the nozzles, wrenches, etc. and also know the location and condition of the hydrants." Good advice, but not enough to prevent the disaster of 1898.

The fire started in the middle of a June night. An excited Chinese shrieked first warning, then a policeman signaled with three pistol shots followed by an alarm from the Marsac mill. It was never determined whether the blaze started in a room or the kitchen of the American Hotel, but a wind snaked flames down Main Street and fire soon threatened to consume the whole town.

Frantic women dashed their children to safety, then seized their belongings, while men blew up their own homes with dynamite in attempts to save their neighbors'. Ashheim's Store was considered a safe repository for valuables, but its tin roof collapsed into a brick shell; the Opera House was gutted; the *Park Record* office blazed with flames feeding on irreplacable documents, and editorial gear was moved to a hillside tent; Park Avenue, Rossie Hill, and Chinatown were demolished. No human lives were lost, but, in less than four hours, more than 200 businesses and homes were destroyed, 500 persons were homeless, scores of livestock were killed, and losses topped $1,000,000. Fire departments from Salt Lake City, Ogden, and Coalville arrived to douse ashes.

Money and supplies poured in from all over the country — shipped free of charge by the Union Pacific and Denver & Rio Grande railroads — churches united in efforts to clothe and feed the destitute, and Mormon farmers from the surrounding area donated dairy products.

Park City was rebuilt from rubble. Within eighteen months, there were more than fifty new businesses and homes, mostly of cheaper and faster frame construction. The *Park Record* glowed, "There is not one camp in a thousand that would have recovered from the blow, but Park City did and with a rapidity that is astonishing."

The lavish homes and ten-story skyscrapers rose in the valley of Salt Lake City, however. Thomas Kearns built a mansion on East South Temple which saw duty in later years as the governor's official residence and now houses the Utah State Historical Society. He also built the Kearns Building on Main Street and the old St. Ann's Orphanage, now a Catholic school.

Colonel William M. Ferry donated twenty-two acres of ground

City Hall stands as rubble after the fire of 1898.

Firemen battle the blaze of 1898.

Welsh, Driscoll and Buck's grocery.

for Westminster College and gave $10,000 towards the construction of Ferry Hall.

David Keith, Ezra Thompson, and Susanna Emery-Holmes added to the architecture of East South Temple, and Main Street expanded with the McCornick, Ness, Keith, Boston, Newhouse, and Walker buildings. All were the products of Park City fortunes.

But twentieth-century innovations were not lacking in Park City. The lavish Dewey Theatre eclipsed the former Opera House by adding a mechanical floor. When the building was to be used for performances, rows of seats were tilted to a convenient angle; when Professor Younger staged a ball or the hall was hired for a prizefight, the floor was laid back flat. The dancing master also added a bowling alley in the basement. Although another show house, the Liberty, tried to compete with special "flicks" and drawings for cash prizes or turkeys, the Dewey's popularity endured to the end. The night its roof collapsed under a record-breaking snow, 300 patrons had departed only hours before.

Mother Urban conducted activities on "the row." The Madam, undaunted by 200 pounds and a wooden leg, paid her fines at monthly raids, contributed to charity, and didn't mingle socially. Parkites realized that her profession kept the bachelor miners from skipping town on payday sprees and made Park City streets safe for any "respectable lady" day or night.

One mayor decided to close the district, however. Mother Urban was incensed; not only would she lose trade, but the mines would suffer from their employees staying down in Salt Lake City three or four days every month. She decided to appeal to one of the company superintendents . . . a bachelor. Within a few days, he used his influence with the city council and twenty-five "seamstresses" were ensconced in the little row of houses.

Prohibition brought new problems. Seventeen bootleggers were arrested and eight soft drink parlors padlocked, yet federal officers named Park City as "the wettest spot in Utah." In one raid, 10 gallons of finished liquor, a 75-gallon still, 500 gallons of mash, and twenty-two 50-gallon barrels were confiscated and burned. Said the newspaper, "the dirty tin tubs into which the dirty booze had been poured were being eaten away by the stuff called whiskey, so the chances are the lives of some of the patrons of whiskey 'merchants' of the city were saved by the timely confiscation. . . ."

They had a roaring time; there were successes and failures, but as with all bonanzas, an end had to come. Mining profits were threatened until, one by one, the mines were forced to close down. Out of work, hit by the depression of the thirties, miners moved to more promising areas and Main Street became a boarded-up, deserted hill inhabited by only a few people cherishing their memories of better days.

Do-ers

THE OLD PARK CITY BUNCH

When a pick and shovel man stumbled onto an outcrop of ore or followed a hunch into a silver vein, his transition from mucker to mogul sprinted quickly. When an investor grubstaked a winner or a mine boss smelled out a few "glory holes," they were destined to become millionaires. More than twenty such men composed the "old Park City bunch" who enjoyed their sudden fortune in true turn-of-the-century fashion.

Their origins varied. They were born in Canada, Nova Scotia, Ireland, Illinois, and Ohio. They were individualistic in temperament and appearance and their ages spanned thirty years. Everything else about them was remarkably similar. One could almost relate one biography and afix it to them all, with only a few minor adjustments.

Our composite Parkite struck out independently in his early teens, orphaned, impoverished, or just plain angry. California was usually his goal and he worked his way there freighting across the plains, ranching, signing on Missouri River steamers as cabin boy, or building railroads. If there was mining along the way, he was in it, and since he had little education beyond common school, he learned about ores and the character of formations in which they were found from observation and experience. Most of all, he discovered that the search for precious metals was the lodestar which would drive him for the rest of his life.

He gambled on scuttlebutt and hefted his pick over the rumor trail. When California slackened, he plunged into the Comstock and when that began to fade, he looked towards Montana, Idaho, and Utah.

Park City's Ontario mine acted as a catalyst to his success. It was the turning point in R. C. Chambers' life because he was the scout sent by financiers Hearst and Haggin to negotiate the original purchase and ended up as the stock-sharing superintendent. It drew David Keith to install the Cornish Pump, Thomas Kearns, John Judge, J. J. Daly, and other future magnates, to dig ore. Friendships developed into partnerships, private prospecting ventures

Susanna Emery-Holmes, one of the old Park City bunch who found their fortunes in Utah silver.

Thomas Kearns, who came to Park City with a pack on his back and left a U.S. senator.

struck Mother Lode, and perserverence soon mushroomed the district out of the one-mine category.

THE KEARNS-KEITH PARTNERSHIP

Tom Kearns signed on at the Ontario, working full shifts and devoting free time to prospecting and studying geology. An Irishman born in Canada, he received a common school education in Nebraska, left as a teen-ager to prospect the Black Hills, then drifted to Tombstone, Arizona, as a teamster and miner. Finally, at the age of twenty-one, he drove a team across the desert to Utah where he worked on D.&.R.G. construction between Salt Lake City and Ogden.

Kearns developed a friendship with David Keith. Keith, born in Nova Scotia, was the youngest of thirteen children. He was orphaned at age fourteen, set out on his own, and was taking contracts for Nova Scotian mines within a few years. At nineteen he sailed to California, heard about the Comstock, and boarded a stage to Nevada. There he took charge of a Chinese work gang for the Central Pacific Railroad during the winter and operated a sawmill in the spring before getting back to what he really wanted to do — mining.

As one of Virginia City's "hot water boys," he made his reputation as a pumpman and foreman for fifteen years. When the Ontario decided to install the Cornish Pump, Keith was sought for the project. He stayed to oversee the No. 3 shaft for eight years before moving to bigger things.

David heard that the Woodside property was yielding some high-grade ore on the eastern tip, so he urged Tom to prospect Treasure Hill just south of Park City on the border of that canyon. The story goes that Tom took three Irish buddies and headed for the hill. They worked several weeks until one evening it was discovered that someone had dropped something back on the trail. Tom and two others conducted a search while the fourth miner waited. When they returned, their friend had a chunk of high-grade ore in his hand. He had sat down to rest, casually chipped at the rock next to him, and happened on a "glory hole."

Tom gave a yelp, vowed secrecy, and chortled, "Let's get a lease on the whole damn hill!"

That "glory hole" was destined to have "no further honor in the future than to be a toilet for pedestrians going up and down the canyon," but Kearns and Keith switched their loyalties to the Woodside workings, contracted to build a 200-foot tunnel, and followed an ore vein into the neighboring, undeveloped Mayflower claim. They leased the Mayflower, and Kearns' gargantuan efforts to prove it resulted in a strike that earned $1,600,000.

As ore bodies at the Woodside and Mayflower became more persistent, they leased additional property, finally hitting another bonanza at the Silver King Mine. (Tom's holdings in this property between 1892 annd 1901 had a market value over $4,000,000.) From then on, they were meshed in dealings ranging from mines, mills, and a rash of lawsuits, to purchase of the *Salt Lake Tribune*. (Keith later sold out his interest to Kearns.) Usually, if one was involved in an enterprise, the other was there for fifty percent.

Tom and David were dissimilar by nature. David was a man of quiet reserve, knowledgeable but not loquacious, congenial yet retiring. Tom made his presence known with gusto. Each was a foil for the other.

It is told that one day, in the temporary absence of the manager, Mr. Keith was barred from his office by a new employee. Without identifying himself, David took a seat and waited patiently until the manager returned and then good-naturedly waved away apologies or explanations.

Stories about Tom are more lively. Some remember the time an anti-Catholic group of Parkites hired a professional boxer from New York to pick a fight in a local Irish bar. Tom stepped in and knocked the pro's head against the brass rail. When the boxer recovered and came down into the mine with four other ruffians to "get" Tom, Kearns said he'd lick them all in a fair fight and told them to step on the cage (which had no sides) and when they were hoitsed up to where they could smell fresh air, he'd take them all on at once. They left him alone.

Thomas Kearns with his wife and daughter in front of their mansion (now the Utah Historical Society).

Sometimes Tom's feisty spirit got him into trouble. When the Terrible Turk threw out a $1,000 challenge for a wrestling match in the Salt Lake Theatre, Tom accepted, pitched the Turk into the orchestra pit, and sent his mother $750 to pay off her mortgage. Thinking he would receive a thankful note, he was mortified to get a scathing letter from his mother scolding him for fighting. For years he kept the letter locked in an old trunk and went to read it privately now and then. He kept his secret guilt so well that his family thought the mysterious trunk must be filled with gold. Upon his death, they opened it and found the letter. Tom had written on the envelope, "But she kept the money."

A cattleman's magazine, *The Kings and Queens of the Range*, praised Thomas Kearns because "his story of success, honor, and fortune (due entirely to his own efforts, and, through almost insuperable difficulties) might well furnish a text for the ambitious youth of today, who craves the world's recognition, but lacks the spirit of achievement."

Keith and Kearns both built mansions a block apart on Salt Lake City's fashionable Brigham Street (now East South Temple). The Kearns residence was elegant with a circular stairwell that swept from a vestibule floor of patterned African and Roman marble to

a gilded dome, three stories above. The stairway's newel posts were French oak and they were surmounted with bronze figures. Walls of the Louis XIV drawing room were hung with silk brocade and those of the banquet hall were tapestry. There was a billiard room, ballroom, a back stairway from attic to cellar, and a bowling alley in the basement.

The unique feature of the Keith home was an octagonal central corridor which was echoed by a smaller well in the ceiling of the first floor. The well was surrounded by columns, arches, and railings and afforded a view to the lower hall from the second story. The second floor ceiling, in turn, held a chipped glass dome set in lead quarries which diffused delicate lighting below. All rooms on the ground floor opened into the octagon and contained *objets d'art* from around the world.

Tom Kearns always championed the little man. At a constitutional convention in 1894, he argued in favor of an amendment on mining taxation intended to help miners: "Are you going to put a tax on that man and sell out what he has put his life to accomplish . . . a claim that he had patented. If you do, gentlemen, you will strike down one of the grandest industries in our new State, and I leave it to your generosity to relieve the poor prospector. Go into the cities and ask capital to go there and invest before the shaft is sunk, it will laugh at you."

He took his crusade into politics when elected United States senator in 1900. But his colleagues in the Senate often mocked him. Not only was he one of the youngest members (thirty-eight), but his background and lack of education drew derision.

The *Salt Lake Tribune* joined in the jibes:

> Kearns will take the place of Senator Wolcott as the best dressed man in the Senate. Kearns has Wolcott's rotund form; he dresses in well-fitting, fashionable clothes, parts his hair mathematically in the middle, and invariably wears a red carnation on the lapel of his coat. But in looks and clothes the resemblance between the two will end, for with Wolcott the Senate will lose one of its finest orators and readiest debators, while oratory, as Kearns himself states it, is not the 'long-suit' of the new Senator from Utah.

Kearns became a good friend of President Theodore Roosevelt, even hosted him as a house guest, but his single term as senator was eventful only for some conservation work in Utah, the opening of the Uintah Indian Reservation, and the raising of Fort Douglas to a regimental post.

Success never made a snob of Senator Kearns. Many squibs in the *Park Record* refer to miners' invitations to parties at the mansion and Tom's equal courtesy to those in silk toppers or flannel shirts. An article scoffed that, while the hoi polloi left on one occasion with bottles of champagne stuffed in their pockets, no Parkite would think of leaving with any liquor he couldn't carry inside himself.

Several scandals plagued the old Park City bunch. One can imagine how the unassuming David Keith must have suffered over the bizarre circumstances that concerned his daughter, Margaret. Margaret, "The Mystery Millionairess of Hollywood," hid her face behind impenetrable veils for forty years. From the age of nineteen on, she perpetrated this self-inflicted seclusion, then staged her eventual suicide so that no one could see her face until she was dead.

It is said that the heiress (her father's estate left her over one million dollars) possessed average good looks but a timid nature that caused her to forego her family's attempts to interest her in society. There were various explanations for her reclusion. Rumors claimed that she had been badly burned or avoided men because of a disappointment in love.

Recently, a Keith descendant was shown a deposition supposedly written by David Keith which proposes another reason. The document states:

Salt Lake City, Utah
April 20, 1898

In the year of 1865, I, David Keith, and my cousin Paul Keith, were rivals for the hand of Manda Summers (sister of Julie Ann Summers). Paul, having won the race, compensated by naming his first son David after me.

David was nine years old when Paul in failing health, and reduced circumstances, begged me to adopt his and Manda's son David (my namesake). Confessing, on his deathbed, that he had tricked Manda into a loveless marriage with him (knowing her love was for me), was in a way recompensed when I made their son David my legal son and heir.

Passing years found me prosperous in mining industries, and David shared with my daughters the advantages a man of wealth could bestow on his children.

Until he reached manhood, David proved a loyal son — he then committed a dishonorable deed, from which my heart has known no forgiveness.

In the home of Alfred Milton and Julie Ann Wilson of Des Moines, Iowa, on the 10th day of April, 1893, a baby girl was born to my youngest daughter, Margaret Keith. As a timid unsophisticated girl of sixteen years, she fell a victim to David Keith's (my adopted son) illicit love.

In protecting the family name, Margaret, believing her baby dead, tried to bury her shame with it, but a veil of darkness shut out all rays of sunlight for her forever.

Sworn to keep the secret of David's and Margaret's unfortunate affair, Alfred Milton and Julie Ann Wilson were appointed guardians of the

David Keith

Margaret Keith

Mary Keith

An elegant home on South Temple was one result of the Keith fortune.

An ornamental skylight adorned the entryway of the Keith home.

illegitimate Keith baby girl, which later became known as Myrtle Mae Wilson.

Only upon the following conditions were the guardians to have full care of said child to raise as their own.

Number 1. In justice to the child, adoption by the guardians or any other person, would not be permitted.

Number 2. Registered at birth as 'Manda Margaret Keith,' the child shall retain same, as her lawful legal name.

Number 3. As the child's birthright, a trust fund of $70,000, in trust with the guardians of said child, shall be used as culture and educational advantages.

Number 4. Margaret Keith's secret (as it was in life) must remain buried with her in death.

Number 5. Not knowing the results of his infidelity, David's past life must remain a closed chapter — to the public.

Number 6. Only after the survival of three years (with complete established identity) shall this confession be given my daughter Margaret's child. In the event the child does not survive my daughter Margaret, this confession *shall not* be given any other person.

The statement concludes by disowning the adopted David and detailing trust agreements and final settlements.

In later years, Margaret, herself, explained her behavior in a note to Charles V. Hatter, a detective she employed. She wrote: "I'm just about as mysterious as an onion. I'm not even a man hater. I like to talk to men because they have brains, but I think the Amazons had the right idea. They made all of the men stay out in the shed with the cattle."

Her suicide could be fodder for a Hitchcock-type thriller, however. At exactly four o'clock one afternoon, her gardner followed a special letter of instruction and placed a call to Dr. J. S. Wood, in Hollywood, and asked him to come to the estate. When the physician arrived, the gardner gave him another letter requesting him to enter the house alone.

Decaying grandeur of the mansion was shadowed by blanketed windows and there was a musty odor. When Dr. Wood entered Miss Keith's darkened bedroom, it reeked with chloroform. He saw "a figure in a black, shroud-like garment lying on a divan, with a great bouquet of flowers at its head and a basket of blossoms at its feet. Over the face was a towel, wrapped in the shape of an ether cone and an empty bottle lay on the floor." Two black cats stirred. Symphonic music swelled from a radio.

A suicide note requested fresh flowers and an orchestra to play classical compositions as the only burial ceremony. The undertaker was instructed, "Please have your lady attendant care for my body. I want to be quite sure that I am dead. Please inject an extra amount of formaldehyde into my body. Cover my hands with

flowers. When I was alive I always wanted plenty of room, so I do not want to be put in a niche." Miss Keith was cremated and her mysterious secret was never revealed.

R. C. CHAMBERS

While the Kearns-Keith partnership profoundly influenced the diggings and doings in Park City, R. C. Chambers' association with George Hearst and J. B. Haggin was also responsible for many developments.

Chambers, like many Parkites, left his Scottish parents in Lexington, Ohio, and joined the flood of teen-agers headed to "make it" in the West. He became part of California's gold rush on Mormon Island in the American River, then settled in Upper Feather River mining camps until 1869. He grubbed in Nevada and as far north as Helena, Montana, finally taking over as manager of Hearst's Webster and Bully Boy mines in southern Utah. When rumors filtered from Park City hills, he was sent to investigate and stayed to open up the Ontario strike and bring it into bonanza.

Upon Chambers' involvement with the Ontario, he formed an agreement with Hearst and Haggin that he would be entitled to one-third of any properties he acquired for their account. This understanding was to seed intrigue on several occasions.

In the late 1870s, when Utah mines were reaching a peak of production, the Union Pacific contrived to force the companies to buy coal at ten dollars per ton from the railroad's fields in Rock Springs, Wyoming. In order to prevent purchase of fuel from closer Coalville, Utah, at two dollars per ton, the number of freight cars to that area was reduced.

Utahns were infuriated. They determined to raise funds for construction of a competing railroad between Coalville and Salt Lake City. On December 27, 1879, the Utah Eastern Railroad Company was incorporated and R. C. Chambers was named to the board of directors.

But the public was unable to acquire enough money through stock sales to build the road, so they appealed to the Ontario Mine, the largest potential railroad customer, for a loan of $186,000. The request was granted and Chambers, superintendent of the Ontario, assumed an additional title as president of Utah Eastern. Mortgage bonds were issued and a spur track reached Park City in December 1880.

The Union Pacific objected to the prospect of competition. In turn, they bought the defunct Summit County Railroad between Coalville and Echo, which had been built eight years earlier. By January 1881, they had paralleled the Utah Eastern line all the way to Park City.

Now, at the time the Ontario loaned the $186,000, J. B. Haggin, the mining company president, persuaded Chambers to give him a bonus of Utah Eastern stock. Chambers complied but neglected to inform the trustees that Haggin was in control. Two years later, Haggin sold his mortgage bonds to Union Pacific and gave them his proxy on the bonus stock. The railroad then shocked the annual stockholder's meeting by announcing that there was a new board, new management, the books were at the Union Pacific home office in Omaha, rolling stock had been transferred to subsidiary lines, and the Utah Eastern railroad was thenceforth closed down. Although coal rates were greatly modified, Union Pacific's Echo and Park City line took over all contracts. The stockholders protested but their railroad remained inactive, was finally sold at auction, and its tracks pulled up in 1887.

Chambers' shenanigans backfired in another instance, however. When Kearns and Keith put their instincts to work on what were to become the great Mayflower and Silver King mines, Chambers wanted in on the deal. He invested a grubstake, but, cautious to avoid complications involving his agreement with Hearst and Haggin, placed the stock in the name of Albion B. Emery, a bookkeeper.

The partners were enjoying profits from the Mayflower before the Silver King ever produced. By that time, they were involved in an apex lawsuit that claimed the Mayflower was shipping ore from the neighboring Northland Mine. The case dragged over several years.

Overnight riches led Emery into a series of bacchanals that resulted in his death from liver and heart failure just at the time the Northland-Mayflower suit reached court. His wife, Susanna Bransford, assumed his partnership and approached Chambers to get her stock. He refused.

Then Chambers made a significant mistake. Called as an expert witness at the trial, he was cross-examined and asked if he was a stockholder in the Mayflower. He was reluctant to reveal his position so answered, "No."

Later, he sued Susanna Emery for $174,717.57, alleging that Emery had purchased one-fifth interest in the Mayflower lease and sold Chambers half of it for $4,000, Emery to hold the interest in trust. He claimed that afterwards, when leases were secured on the Silver King, their agreement was extended to cover that stock. But Susanna told the judge that the grubstake was only a loan which had already been repaid. The court decided in her favor and in a short time, her 150,000 shares were valued at $80 each. (By 1902, she was worth $100,000,000, with a monthly income around $50,000.) Susanna Bransford Emery Holmes Delitch Engalicheff was launched as another Park City bonanza.

Albion B. Emery was married to the Silver Queen when she was just plain Susie.

Susie and Albion Emery's home in Park City was comfortable but humble in comparison with her later palace. Susie's adopted daughter Grace is sitting in the background.

Amelia's Palace (or Gardo House), home of the Silver Queen.

Above: A separate art gallery was attached to the Palace.

Below: Art objects from around the world — and always fresh flowers — in the salon of the Silver Queen's Palace.

THE SILVER QUEEN

In Park City Susanna was just plain Susie, but from Paris to Moscow and New York, from London to Peking and San Francisco, she was "Utah's Silver Queen" and she played the role with éclat. She trailed the names of four husbands like an ermine stole and blazoned her motto, "Why live if you can't enjoy yourself?" on society around the world.

Susanna, born in Richmond, Missouri, and reared in the California gold country, came to Park City for a visit with friends and stayed to marry Emery, the local postmaster, secretary of several infant mining companies, and, later, Speaker of the House in Utah's Thirty-first (and last) Territorial Legislature.

Shortly after their marriage, Emery wrote her parents, "The smile on Susie's face is daily becoming broader and my intelligent grin has become perpetual, so no good reason why this agreeable state of affairs should not continue.... We intend that it shall." Their happiness endured, but was fated to a brief ten years. His alcoholic excesses killed him on the return trip from a vacation to Hawaii.

When Susie was left to support herself and eight-year old Grace, her business acumen attracted praise. In an interview with the Washington *World*, she said, "As the trustee of my daughter's fortune, I am compelled to manage all the mines and railroad interests of my late husband. Some of my friends have said no other woman is engaged in such large mining enterprises or has personal supervision over such immense financial interests."

When Grace came of age, another article reported her estate close to $2,000,000 "which leaves no doubt as to her being the wealthiest young woman in the intermountain West."

Shortly after her eighteenth birthday, Grace married one of Susie's nephews. From that time, the mother-daughter relationship faltered and when Grace died, childless, at the age of thirty-one, Susie sued for her estate and lost.

With passing time, perhaps Susie remembered the words Emery had once written in her autograph book:

My Friend
Life is too short for any vain regretting;
Let dead delight bury its dead, I say
And let us go upon our way forgetting
The joys, and sorrows, of each yesterday.

Widowhood palled after two years and the slender, brown-haired charmer married Colonel Edwin F. Holmes, a wealthy Chicagoan with lumber and mining interests in Idaho and Utah. He came to

Susanna Egeria Bransford Emery-Holmes Delitch Engalicheff, "Utah's Silver Queen," about 1902–4.

Jewels and elegant clothes were Susanna's trademark.

A table set for guests in the Palace.

check his holdings in Park City's Anchor Mine and ended up saying "I do" at the Waldorf-Astoria Hotel in New York City. Susanna Emery-Holmes then became the "Silver Queen."

She dressed! News sheets touted, "Some of her superb gowns are made in Paris, but, although the brilliant French capital leads the world in styles, she attaches no special importance to the place where her dresses are bought or made, buying what pleases her most whenever she sees it."

Her wardrobe included rare laces, jeweled jackets, heavy satins and brocades, and treasures such as a $10,000 circular charm of platinum and diamonds. Seven trunkloads of her clothes were later donated to the Daughters of the Utah Pioneers Museum in Salt Lake City.

She entertained! Luncheons in Washington, D.C., for the German Kaiser's brother; musicales for select audiences to preview Salt Lake Theatre's coming attractions such as Alberto Jonas, the Spanish pianist; elaborate dinner parties featuring fortune-tellers and string quartets; and a wedding for her sister that tickled Salt Lake City tongues with tales of flowers and trousseau worth $100,000 and a gift to the bride of $50,000 in silver mining stock.

The *New York Herald* headlined her battle for social supremacy with Thomas F. Walsh, the reputed prince of hosts, and Susanna scribbled "funny" in her scrapbook beside a clipping from *Goodwin's Weekly* which stated, "Her dinners are dreams and her invitations are a great privilege to the fortunate ones, but she does not entertain enough to lead." Her scrapbook also displays a personal invitation to the coronation of Edward VII of England and she claimed to have once danced with the Prince of Wales — "And what a dance I gave him!"

Every queen must have a palace and Holmes saw to it that Susanna did. He gifted her with the Amelia Palace (also known as the Gardo House), an imposing Gothic structure Brigham Young built for his favorite wife. Twenty decorators from Chicago's Marshall Field Company conspired for a lavish Christmastime opening, and Salt Lake society was dazzled by a Louis XVI salon with walls of old rose brocade satin, ivory enameled woodwork, reseda green carpet, and draperies and upholstery of Persian silks and fine skins. Paintings, porcelains, and world-wide bric-a-brac clustered on the walls and filled gold-carved cabinets.

A Gothic dining room, papered with gold leaf overlaid with lustered green and blue, was furnished in carved Belgian oak and carpeted rich red.

Midway between the two rooms, an electric Tiffany fountain could be converted into a huge centerpiece by arranging feasting tables on either side.

There was a broad, curved stairway, a secluded, book-lined

Prince Engalicheff,
Susanna's last husband.

Susanna with an unidentified
husband on a tour of the Orient.

Susanna on one of her many tours.

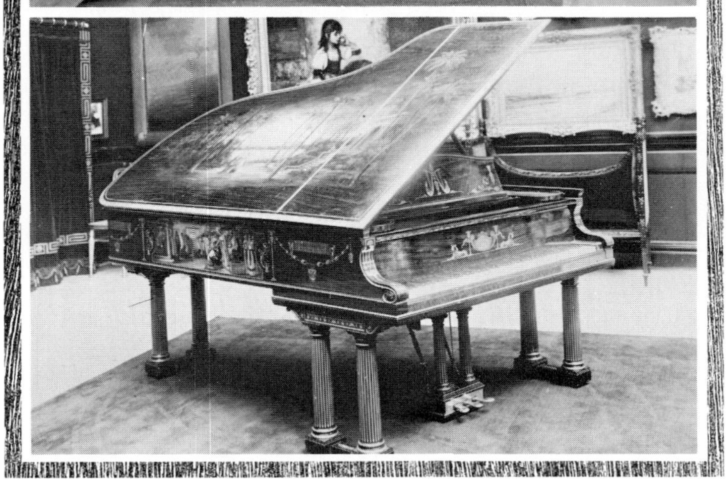

Above: With Col. Holmes on a porch of the Palace.
Below: Musicales were often held in the art gallery.

Susanna had a collection of Indian photographs
in the second floor hallway.

Utah's Silver Queen.

tower, a music chamber, an Oriental room, an art gallery — almost fifty rooms, each elaborate, each distinctive. It is said that when Susie entertained, she made grand entrances down the stairway while her full-length portrait was unveiled in the foyer.

The palace doesn't stand any longer. It was used as Red Cross headquarters during World War I and when Holmes and Susanna moved to Southern California in 1924, it was sold back to the Mormon Church. The building was demolished in 1926 to make way for a new Federal Reserve Bank, which was later sold back to the church and now houses church offices.

After twenty-eight years of marriage, Susanna was widowed again, but it wasn't long before she met a broad-smiling Serbian doctor, Radovan N. Delitch, and added another to her list of names. Within two years his dashing ways began to embarrass her and she became impatient with his jealousy. She bade him goodbye in the divorce court and he hanged himself in his cabin aboard an Atlantic liner.

It only took Susanna a year to find her Russian prince; she then became "H. R. H. Princess Engalicheff" on hotel registers. They were twice married in New York, in civil and Greek Orthodox ceremonies, but one chronicler tattled that the "mustachioed descendant of the Tartar warrior Genghis Khan" went through the intricate gestures of the Byzantine ceremony well except that "the gold crown which, obligatorily, was placed on his head was a bum fit and slipped down over the Prince's nose."

It might have been headlines such as, "Now the Romantic 'Silver Queen'... Aged 74... May Ditch Her Russian Prince to Marry a Boy!" or it might have been Engalicheff's infirmities, but, within two years, Susanna was widowed again. The "boy" became her business manager and her reputation waxed as *grande dame*, business woman, cosmopolite, and breezy westerner.

In 1938, Susanna returned to Utah for the first time in years. She and her business manager stepped from a suite of D.&.R.G. drawing rooms complaining that Florida was too cold, New York too hot, Southern California much too shadow-crowded with sad memories of Holmes, Newport banal, and Maine monotonous. She was sated by the glitter of European capitals and hoped never to see them again.

A reporter asked if she had any marriage plans and she replied, "I am unfortunate in keeping husbands." It is said that she later lunched with Thomas Kearns' widow and expressed surprise that she still wore the senator's wedding ring. When Mrs. Kearns asked whatever else she would do with it, Susanna chuckled, "I hang mine in the bathroom, like grapes!"

The era of the Silver Queen ended in 1942 when Susanna was 83. As in life, death brought many stories about her. While the *Salt*

Lake Tribune reported that "she died quietly while stopping at the Putnam Inn, Norwalk, Conn.," there were those who felt that she committed suicide; some who suspected foul play.

"Why live if you can't enjoy yourself?" Susanna had asked. She always said that when she ran out of money she would do away with herself. At her death, her entire fortune had dwindled to little more than she used to realize in one month. When her body was returned to Salt Lake City, in keeping with her request to be interred beside her first husband, relatives had to pay for the funeral. The bulk value of her $65,918 estate came from 15,020 shares of capital stock of the Silver King Coalition Mines Company. The only one to profit from her depleted fortune was the business manager, who got all but 5,000 shares of the stock and some apartment house property at State Street and North Temple that developed enough resale value to set him up for life in Sicily.

Just three weeks before she died, she had moved from New York City to the Berkshire Hills and, at the time of her death, she was en route to visit friends in Virginia. Those who suspected foul play argued that such plans would not indicate suicide and, besides, it would be out of character for one who had flourished through life to end it with so little fanfare. Also, just before leaving on her trip, she had written her sister a letter saying that she was afraid.

But there are those who wonder if she was just afraid of being broke and old, with nothing on the horizon. Perhaps she did just die "quietly while stopping at the Putnam Inn."

Willis Adams left many paintings and photographs of early Park City.

WILLIS A. ADAMS

Most early Parkites followed the silver lodestar. William McCornick laid the foundation for his fortune on the Comstock, then came to the Park to invest further in mining and establish "the largest financial institution of its kind between the Missouri River and the Pacific."

W. "Mont" and E. P. Ferry left thriving businesses in Michigan to dig Utah hills. John Judge, J. J. Daly, Richard MacKintosh, Jim Ivers, David McLaughlin, J. R. Murdock, and Ezra Thompson were others in the "old Park City bunch."

But it was forests and hidden lakes, mountains and flowered meadows that attracted Willis A. Adams, and what he loved the most he preserved with paint or pencil.

Adams' love of nature emerged in delicate rendering of line and luminescent colors. Three of his oil paintings were exhibited at the St. Louis World's Fair and hundreds of oils, watercolors, pastels, and drawings hang in private collections.

He was a photographer and poet as well as a painter, and combined artistic talents with the inventiveness and practicality born

Old-time do-ers dressed elegantly for a mountain outing.

of frontier farming origins in Indiana. He constructed many tools to help in his photographic trade, among them a lighting apparatus for taking photographs at night or inside the mines. If a sudden urge to hike caught him without a canvas or camera, he sketched on slabs of sandstone or, one time, on the broken spade end of a shovel.

The large store of photographs might well be Adams' greatest legacy. His pictures show Park City's pre-fire days, the surrounding countryside, closeups of the Opera House, Ashheim's Store, the Cornish Pump, mines and their boardinghouses.

Unlike the mining magnates, Adams lived in Park City for fifty years. When he did move to Salt Lake City, it was to a cobblestone Spanish-style bungalow which stands yet in Fairmont Park.

A view of Park City at the turn of the century.

Park City...Now

Harry Harpster

Park City ruggers provide summer excitement.

Doings

It was there all the time, but no one saw it. Winter sunshine, silvered, snow-packed slopes; bus loads of Girl Scouts chugged up the canyon bound for camp; Sunday drivers cruised hilly streets to ogle ramshackle houses with boarded windows and toy-town wooden stairways climbing into fields of neglected hollyhocks. Nothing was going on, but Park City was a hard place to ignore.

By the late 1950s there were fewer than 200 men working the mines. Lead and zinc prices were depressed. Yet the United Park City Mines Company didn't want to disregard the camp until economic conditions would permit large underground developments that were on their books. They needed some method of providing income for the firm during the difficult period.

They made an assay. Along with underground holdings, the company owned surface and water rights to 11,000 acres which included roads, power lines, and real estate. The place had atmosphere, there was ski terrain, rail transportation, projected freeways. It was a natural for a recreational area they decided. All that was needed was the wherewithal and Park City's surface was certain to prove as promising as its heart.

Funding came. An Area Redevelopment Loan of $1,232,000 was granted by the Department of Commerce. Private sources of Anaconda Company, American Smelting and Refining Company, Park City Recreation Committee, and United Park City Mines Company matched with $1,894,860.

In 1963, blasts of dynamite and a rainy Maytime parade announced the birth of the Treasure Mountains Resort. People started coming back; real estate boomed; light was let into the houses. Main Street swelled with false fronts and Victorian plush, and Park City divided like an amoeba, part old-time mining town, part Alpine village — but each important to the whole of Utah's newest year-round recreational area.

The beginnings were enthusiastic and smacked of whimsy. Twentieth-century go-go girls jiggled in Gay Nineties saloons, rock bands blasted under Tiffany chandeliers, a mural placed "Utah's Silver Queen" in a Mexican setting, and there was a hotel with a French name and Polynesian decor. Main Street was a gap-toothed grin of weathered ruins leaning against gilded façades standing

The new Racquet Club offers year-around tennis.

A rodeoer is about to bite the dust.

Wayne Putnam is a favorite Demolition Derby daredevil.

Skateboard contests.

next to unattached door-stoops leading to weed-filled vacant lots. Park Avenue's eighty-year old churches were within chiming distance of A-frame vacation homes.

"The longest aerial tramway in North America" swept skiers to the mountaintops and "America's only underground ski lift" rattled them in a mine tunnel train to the old Silver King hoist. Treasure Mountain cradled a museum deep in its bosom and a nine-hole golf course at its feet. (The Mine Tunnel Train no longer carries skiers and the golf course has expanded to eighteen holes.)

It was like the prospecting times in some ways, but now it was businessmen, not miners, who invested in their dreams, many to find a "glory hole," others to fail. They came and went for eight years with grand openings and closeout sales. The population revolved around stubborn optimism.

Then, in 1971, Royal Street Development Co. purchased the resort through a subsidiary, Greater Park City Co. For the first ten years, the project would exceed one hundred million dollars. Their master plan envisioned condominiums, homesites, and shopping centers in adjacent cluster communities and a complex of interconnecting lifts and runs aimed at a ski capacity approaching 19,000 persons. There would be hotels, restaurants, boutiques and increased facilities for golf, tennis, swimming, water sports and cultural events. They lost no time in pinning cash and action to their promises.

July 4th parade.

Stein Eriksen, Olympic Gold medalist, winner of three World Championships and noted "skicrobatic," was brought to Utah as the resort's Director of Skiing. Stein joined other deep powder experts in shuttling to high peaks in Hosking helicopters, touring cross country or streaking down Payday, the longest night-skiing slope in the United States.

Capacity of the mountain for skiers was expanded with six additional lifts and miles of new runs. The need for better accommodations for seasonal crowds and conventions was met with construction of 250 condominium units, a central plaza with shops and restaurants and development of 200 homesites.

A rodeo contestant.

But economic and climatic conditions waged a counter attack. The more the resort expanded, the more it became buried in debt due to high interest rates, a slackening of real estate sales and a general national recession. Coup de gras was the green Christmas of 1974, with a dearth of snow that cut that ski season virtually in half. By mid-1975, the Greater Park City Co. disqualified itself from the race. Stockholders cancelled the company's debts in return for real estate mortgages, trust deeds and notes. United Park City Mines Co. cancelled accrued interest and deferred payments of the principal. The 132 newly built Park Avenue con-

An eighteen-hole golf course attracts many locals and visitors.

dominiums were whisked away in a real estate bargain sale that was sold out in three days and Alpine Meadows of Tahoe, Inc. bought the resort.

Alpine Meadows put all their money on the mountain. Skiing was their thing. They divested themselves of food concessions, eschewed lodging interests, and laid $2.4 million dollars the first two years into refurbishing existing equipment, constructing two new double chairlifts and opening Jupiter Bowl for expert powder skiing.

Now, the big bell behind the Resort Center gongs twice daily, at 9:45 a.m. and 1:45 p.m., to gather students for the newly reorganized ski school. Instructors are members of the Professional Ski Instructors of America and private or class lessons are arranged by appointment.

The skiers as well as the miners are wearing hard hats these days and Parkites get their share of competition with NASTAR races every Saturday, the Lowell Thomas Classic the end of January, the *Salt Lake Tribune* Classic for junior champions in March and the Media Medley for radio and television personalities.

Since 1973, Park City has housed the national headquarters for the U. S. Olympic Ski Team. One of the most spectacular races is the U. S. Ski Team Invitational, which was initiated in April, 1975. Top Nordic and Alpine competitors on the national team join forces with celebrities and local skiers for a weekend of races and aprés ski festivities. The function is becoming one of the major fund raising events of the Olympic team.

There's cross country skiing, too, and tourers can reach the snow-laced trees and unbroken trails of outlying canyons and valleys within minutes. Many trails retrace abandoned mining or logging roads and weathered remnants of former diggings are often passed. Novices can perfect their glide step on flat tracks at the golf course with instruction from the White Pine Touring Co., based at the Timber Haus sport shop.

Winter worshippers almost reach a point of frenzy over mid-February's deep powder snow and ideal conditions. They celebrate then with the annual Snow Sculpture Contest which transforms the golf course into an art gallery of mammoth figures molded from ice and snow.

There's skiing down the road at Snyderville, too, where Park West has whittled 28 runs into 32 miles of another mountain. Their ribbon was cut in 1968 when they opened lifts, a ski school and day lodge as the first phase of an extended building program. The area's new owners now boast five double chairlifts, tennis courts, a swimming pool and a condominium village within walking distance of the slopes.

Many baseball teams compete in Park City.

Harry Harpster

Harry Harpster

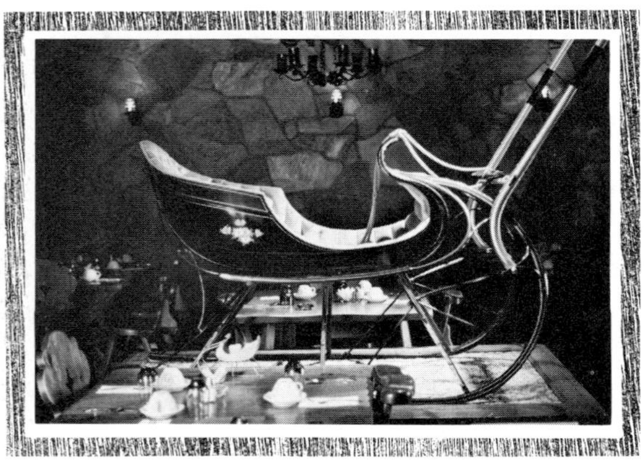

Harry Harpster

Many charming restaurants contribute to Park City's atmosphere.

Park West is the hangout for the "white hat boys" and everything is in keeping with a western theme. Runs are tagged such names as Sourdough, Slaughter House, Ambush and Six Shooter and the Rawhide Depot is where one goes for on-the-mountain refreshments. After skiing, slatsters make tracks to the Barn at the resort center for flicks, beer, food and the attractions of a private club.

Golf is a major warm weather feature in Park City. Duffers and low handicappers, alike, are challenged by the championship eighteen-hole course which skirts the Three Kings and Park Avenue condominiums. Players may sharpen their skills with professional instruction or on practice greens and the pro-shop provides cart and club rentals.

The scope of summer activity has been broadened with construction of tennis courts and swimming pools. There are two outdoor courts available to the public at the golf course and the Park City Racquet Club's four indoor, and seven outdoor courts and swimming pool are open to members and guests. Persons renting condominiums owned by club members are entitled to use the facilities at member's rates. The city park, north of town, keeps three outdoor courts busy and several hotels have swimming pools for their guests.

An exhilarating new summer treat is the Alpine Slide, a 3,000 foot-long winding track down the Payday ski hill. Originating in Germany, the slide is traveled on small sleds controlled by each individual operator. One may choose to ride slowly on a set of runners or push a control level to substitute rubber wheels that will speed the sled up to 25 miles per hour. Two parallel tracks weave through the trees and traverse a vertical drop of 550 feet.

Hikers and rock-hounds track through the hills to seek the solitude of unexpected lakes and wildflowers or sift through deserted mine dumps for pyrites, peacock iron, dendrites, galena and zinc. Some dig around dilapidated boardinghouses for ore testing cups or an occasional Chinese wine jug. Others hoist backpacks and roam a wider range. Many simply ride the Gondola to picnic and enjoy free badminton and horseshoe facilities at the Summit.

Prior to the autumn of 1970, most Parkites were unaware that there was a game called Rugby. The Park City Rugby Club is now a source of civic pride. According to rugger, Dave Mueller, "We all ski. We just play rugby in the off-season. But the trouble with that is that a few of us have gotten to appreciate rugby so much that we can't make up our minds *which* is the off season."

At least one local game is played each weekend throughout the summer. Visiting opponents have come from such places as Colorado, Texas, Virginia and Pennsylvania, and the Park rug-

Harry Harpster

Harry Harpster

Melodrama brings villains to the Opera House on weekends.

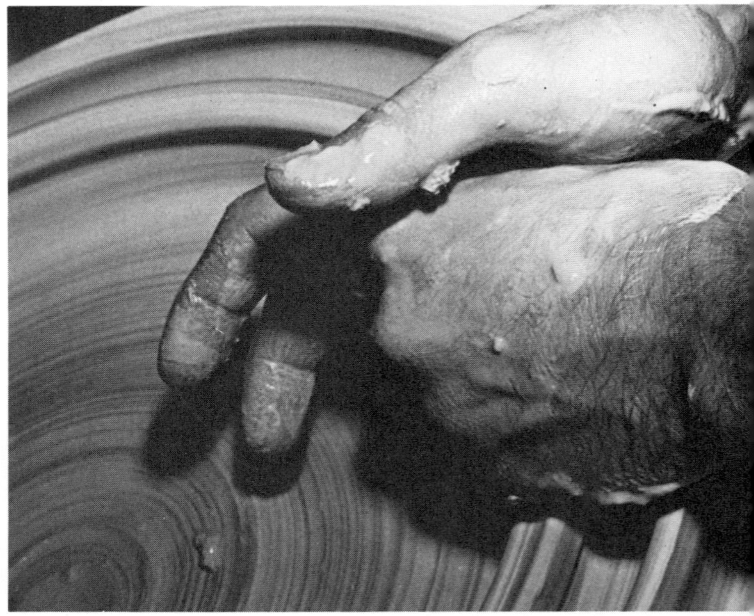

The new Kimball Art Center exhibits work of many artists.

The annual Park City Art Festival attracts huge crowds every August.

gers have also traveled out of state. Not to be outdone, Park City women have also organized a rugby team.

An unbroken chain of parades and celebrations has enlivened every Fourth of July and Labor Day of the town's existence. Kiddies still scramble after greased pigs and shinny up slick poles, miners compete in mucking and drilling contests, and tourists join in it all from the opening blast at sunrise until dark.

The summer season is officially launched the first two weekends of June with an outdoor Jazz Festival and the annual Park City Art Festival, held in August, always accounts for a population explosion. The event attracts over 70,000 spectators who meander among sidewalk exhibits of painting, sculpture, glass blowing, wood carving and other works shown by artists from many western states. Starting as a weekend event, the Art Festival has grown to a full ten days with additional concerts, workshops and theatricals as well as the Main Street exhibit.

Year around cultural and educational programs take place at the Kimball Art Center, which was founded by Bill Kimball (descendant of the old-time livery stable Kimball) and was built on the site of Snyder's boardinghouse (where Park City got its name). Destined to add its own bit of history to the community, the Art Center offers art exhibitions, music and dance concerts, lectures, films, artistic and technical workshops and courses in photography, ceramics, weaving, painting and other art forms.

Dazzling yellow ore cars fill with sightseers three times every day during summer (weekends, in winter) for excursions to the Silver King Mining Museum. The half-hour, three mile Mine Tunnel Train ride is the only one of its kind in America. Illuminated cars rumble through the Spiro tunnel and passengers get a first-hand look at the dripping, rocky underground network of an operating mine. At one point, all lights are extinguished for a few minutes to demonstrate total darkness.

Life-like wax "miners" work diggings in a large central cavern at the museum, where visitors inspect ore carts and stables down track and climb ladders to a timbered stope. Temperatures are in the mid-40's year around, so sweaters are suggested.

A topside view of the Silver King Mine unfolds on the Gondola ride at the resort center. The two and one-half mile tramway glides above the old bunkhouses, boardinghouse, concentrating ponds, ore dumps, hoist house and mill. Abandoned buildings and equipment of other deserted mining giants are visible in the distance and draw an impressive picture of the number and scope of mining endeavors which flourished at the turn of the century.

Dusk drowses over Main Street in a limbo between daytime activities and the time when dancing white-hot globes and bursts of blue, yellow and red neon beckon to evening celebrants. They

The Alpine Slide is a popular new summer activity at the resort center.

Condominiums range from modern redwood to turn-of-the century Victorian style.

come to dine, to party at saloons and to escape from present reality into Park City's lively past.

Like everything else in this mining town turned resort, the restaurants are footed on shifting whims. Names and managements change; a few enjoy relative permanancy, more lead a transient existence. But there are charcoal-broiled steaks and lobsters, buffets of salad fixings, crepes, fondues, Mexican and German delicacies, as well as hamburgers and pizza. There are dining rooms simulating Nineteenth century railroad palace cars, one in an old coal and lumber store, a cellar bistro with tables suspended from heavy chains, plush Victorian salons, coffee shops and cafeterias. Reservations are suggested for major establishments as some are closed between seasons or open on weekends only.

Utah's liquor laws dictate bringing your own bottle, but thirst quenchers are readily available, whether you "brown bag it" at a State Liquor Store or order an evening's supply of "mini bottles" or wine from authorized restaurant outlets. Most lounges serve 3.2 beer. A few private clubs collect nominal membership fees for the privilege of over-the-bar cocktails.

Saloon hopping is a favorite form of nightlife in Park City. Celebrants join impromptu sing-alongs, listen to guitarists and folk singers and chuckle at old-time flicks or movies of themselves on the ski hill. They boogie with the "big beat," stomp to bluegrass, savor a jazz concert and click pool balls over schooners of brew.

At the Silver Wheel Theatre, weekend audiences hiss and cheer as actors unfold melodramas in the red velvet atmosphere reminiscent of bygone Opera House days.

Wood sculptors, artists, leather craftsmen, potters, jewelers and silversmiths; gift shops, boutiques, book shops, antique dealers, sports outfitters and old-time country stores; saloons, pool halls, ice cream parlors, fudge kitchens, plain and fancy eateries — all are grouted into the mosaic of Main Street. And right there with them are authentic buildings that date back to early days.

Just below Main Street is the Silver King Coalition Mines building, a Park City trademark constructed in 1889. The old tramway cable towers disappear up the mountain to the West.

At the foot of Main Street is the Union Pacific depot, dated 1886. Its freight room walls are scarred by miners' autographs, graffiti and doodles burned or painted on the rough wood. The railroad tracks, laid in 1881, are still usable.

Both of these structures are undergoing renovation as The Depot, which will contain a mining museum, railroad museum and restaurants and be served by an interconnecting trolley line bordered by restored railroad cars used as motel units.

Up the street is City Hall, built in 1885. It houses the only territorial prison still in use and there is a dungeon, complete with leg-irons and solitary confinement cells, in the basement.

The First National Bank Building (now the Board of Education and a gift shop) is another survivor of the big fire of 1898.

The *Park Record* has officed several places since its founding in 1880 and is Utah's oldest weekly newspaper of continuous publication.

Looking west to Park Avenue, one sees St. Mary's of the Assumption church. Dedicated in 1884, it is the oldest Roman Catholic edifice in the state.

A short walking tour reveals many churches, schools, residences and other remnants of history tucked among the hillside stairways. A visit to the two cemetaries west and north of town is also of interest.

But equally as exciting as the past is Park City's future — the Deer Valley Resort Area, a Royal Street project that envisions three separate communities, an eighteen-hole golf course and over twenty lifts to accommodate 9,000 skiers; an equestrian center; the Park City Cultural Center, hotels and shopping complexes.

Park City's new life gestated with a single ski resort, but has matured into a vital, permanant citizenry growing with expanding job opportunities and adding toddlers, teenagers and young-marrieds to oldsters on front porches watching their changing town with curiosity, amusement and a twinge of resentment.

Now, behind the neon facade moves a burgeoning community. A lady on Park Avenue sets herself to old-time tasks of quilting or making soap; a young man fires a torch to sculpt metal; a volunteer guild is formed to boost the proud new Art Center; long-haired newcomers join grizzled natives for meetings of the City Commission or the Fourth of July Parade Committee.

So it all goes together — past and present build upon each other. Old and new, ALL are Park City.

Resumption of mining necessitated the new Ontario mill and concentrator.

Old Time miners are back on the job at the Ontario.

Miners in the hoist bound for underground.

Diggings

Like moles, the miners picked their way through the debris of abandoned ore cars, sinking-buckets, rock slabs and rotten timber in the slimy No. 3 shaft of the deserted Ontario Mine. Junk that oldtimers had dumped below the 1,700-foot level to bulkhead the caving shaft was hoisted to the surface, water from lower levels was sucked out by modern pumps having four times the capacity of the historic Cornish Pump and tunnel walls were timbered, wired and bolted for support. New ore bodies were detected, drifts and stoping areas were prepared. They were reviving jokes about the Tommy Knockers and warning each other to watch out for the spectral "Man in the Yellow Slicker." The dormant years were over and they were breathing life back into Park City's first producing mine.

Mining in the Park City District had virtually collapsed following World War II. Depressed metal prices, skyrocketing production costs and prolonged labor disputes had shrivelled the hundreds of mines that flourished during the heyday down to two operating companies in the early 1950's: the United Park City Mines Co., which was a consolidation of the Silver King and Park Utah, and the New Park Mining Co., which leased the Mayflower Mine to Hecla. Ore shipments were sparce and work was mostly exploratory.

Then, in the spring of 1970, the Anaconda Co. and ASARCO formed Park City Ventures and leased all mineral rights, property and equipment connected with the United Park City Mines Co.'s mining operations. They resumed work in the No. 6 shaft, renovated the old No. 3 with steel and concrete, started to drive 33,358 feet of drifts, cuts and raises, drilled almost 153,000 feet of development holes and pumped out the ubiquitous underground water at the rate of 4,500 gallons per minute.

Much travel underground is via narrow planks over rushing water.

Within three years, the shaft was completed and a new hoisting plant, including headframe, production and auxiliary service hoists, an air compressor and hoist building, was installed. Then, construction began on a $17 million flotation concentrator designed to treat 750 tons of crude ore per day, and a surface plant with warehouse, shop, offices, change room and garage. All mining and milling operations would be conducted from this one facility, which was painted desert gold to blend with the environment.

Production started in April, 1975. By the end of the year, Park City Ventures had mined 94,483 tons of ore and shipped almost 20,000 tons of dry concentrate bearing 66 percent lead and 55 percent zinc. Silver content was assayed at 3.6 ounces per ton and there were small quantities of gold and cadmium. The following year almost doubled the tonnage mined and they reported average grades of lead at 68 percent and zinc at 55 percent.

Anaconda and ASARCO divided concentrates on a 60-40 basis. Shipments were trucked down into Park City to a Union Pacific loading station, where the concentrates were sent to custom smelters in Idaho, Montana and Oklahoma.

The No. 6 shaft, or "East Flank" near Keetley, accounts for almost 60 percent of the Ontario's ore production, with No. 3 shaft, the "West Flank," taking the balance. Diamond drilling and other testing is being conducted in surrounding areas of known mineral bearing structures.

The Park City District encompasses numerous ore bodies in lode deposits covering an area of approximately 20 square miles. Mining is undertaken from the surface down to about the 2,400-foot level. It is a difficult district to mine due to an abundance of underground water and great differences in the types and forms of ore bodies, which vary according to geologic environment, host rocks and structural control.

From the original discovery until the resumption of active mining in 1975, the district has produced approximately 20 million tons of ore, averaging 13 ounces of silver, 0.07 ounces of gold, 8 percent lead, 4 percent zinc and 0.4 percent copper per ton, and gross metal values have exceeded $2.5 billion. It is one of the few billion dollar districts in North America.

But there are no latter-day millionaries rising from the Ontario's ashes these days. Financial results are directly affected by the metal markets, which reflect production, consumption and world-wide stocks. Despite the promise of great potential, Park City Ventures' annual statement still shows almost $3 million in the red.

The current 380 employees are not the boardinghouse bachelors and roving professionals of the olden days. Most of the miners are married citizens of Heber City or Kamas. Many are unskilled as miners and must take on-the-job training courses provided by the company.

Two main shifts and one graveyard shift for resupplying the workings put the mine on a 24-hour basis. Miners alternate the 7:30 a.m. - 3:30 p.m. and 4:30 p.m. - 12:30 a.m. shifts every two weeks and they work six days every other week. Those who do drift, stope and raise mining work on contract, receiving $X.00 per foot for breaking rock, timbering, piping track with air and water lines and mucking. Motormen, helpers, electricians, mechanics and others are paid union scale.

It used to be considered unlucky to have women in the mine. Today, women take full shifts underground operating hoists, pumping sand into mined-out stopes, surveying, making primer for blasting and even breaking rock.

As a rule, the women have been accepted on an equal basis with male miners, however, it is said that one fellow, after having been very cooperative with his female partner on her first day, came to work the next day and refused to team with her. The foreman, figuring the man had had a little static from his wife the night before, promptly switched the young lady to another mining partner.

Shelley Christiansen was the first woman to actually break rock in the Ontario. Shelley had heard about mining all her life from her grandfather, father and brothers, who had all worked Park City mines, so she decided to try her hand at it. She started out by working in a sand-pumping station where she merely turned on a machine and sat all day to monitor its operation.

Then one day, after she had worked there a month, her foreman surprised her by asking if she wanted to enroll in a training course to become a contract miner. No other woman had attempted this, but being a plucky lass, Shelley agreed. Next thing she knew, she was on the business end of a jackleg drill.

She learned fast. A five-foot-five, 130-pounder, she had to have someone hold the six-foot steel until she collared the bit into the rock, but then she was off and running — compressed air and water whooshing through the hoses, the spinning bit screaming into the hole, the 70-pound drill jigging as she leaned her full weight against it.

All went well until one day when she gave the drill too much air. Her instructor had just turned to leave her to her work when she gave it the air and the jackleg started to jog across the floor. She leaned harder on the drill. The jackleg took another leap backward. The air and water hoses gyrated. The jackleg bolted and Shelley was pinned against the wall with the drill going full blast. She took a bit of joshing from her colleagues but they quit teasing six weeks later when she was working with her partner in her own stope and mined enough ore in two pay periods to gross the top bonus.

"Hard-boiled hats," safety goggles, hard-toed boots and yellow rubber overalls and slickers are worn on the job by the miners as protection. When they go to work, their first stop is the change room, where they collect their clothes and equipment, which have been hung to dry on hooks and in baskets dangling from long

Shelley Christiansen is the only woman to break rock at the Ontario.

Miners rest and have their lunch in a "dog house."

Track bikes are still used by surveyors and foremen.

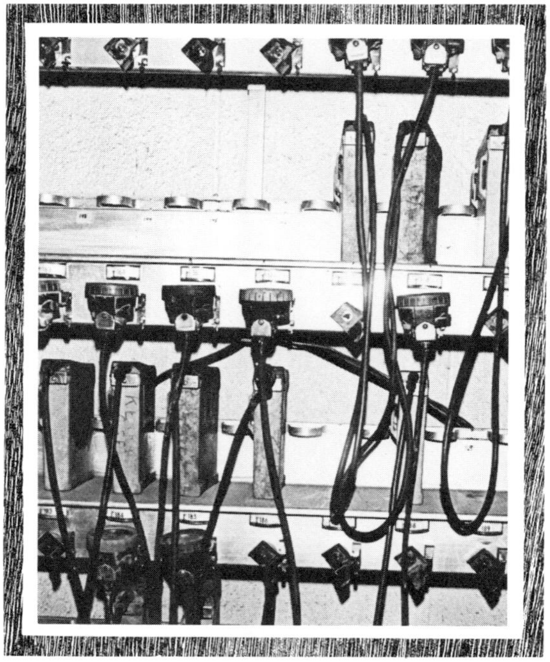

Batteries for head-lamps recharge.

An underground office keeps contact with the surface.

Miners hang their clothes from the ceiling in the Change Room.

chains attached to the ceiling. They pull down their gear, don their mining togs and then head for the hoist and "cages" that will carry them to their workings.

The hoist is operated by a control person who sits on an elevated platform before a large panel of dials and levers which activate huge cable drums to raise and lower the cage. At each level of the mine, there is a call bottle which the miners pull to ring coded signals to the operator as indication of which level they are at and whether they want to go up or down.

The No. 6 shaft hoist is still moved by a rare flat-cable, consisting of approximately thirty strands of cable woven together to resemble a coat of mail. The fifty-year old hoist is one of four or five flat-cable types in the world that is still in use.

No longer horse drawn, the "man trains" are pulled by battery powered locomotives which carry the miners through the famous old No. 2 drain tunnel to their work positions. It takes approximately an hour each trip to bring crews from the surface to their stations.

Underground, there is total darkness, except for electric lights in the hoist room, level stations, compresser room or storage areas. A pinpoint of light from a battery-operated head-lamp is the miner's only guide as he works.

Miners don't have to rely on flickering candles to detect carbon monoxide or smoke these days. A stench alarm system can be activated to release a warning odor throughout the region. Along with a head-light battery on his belt, the miner carries a "self rescuer." Upon detecting the stench alarm, he removes the rescuer from its cannister and utilizes a mouthpiece that converts carbon monoxide to carbon dioxide as he escapes up ladders or through emergency exits.

Separate lines for air, water, power and telephones are strung throughout the mine and special air doors close off tunnels at strategic points to force air that is fanned from the surface down to lower levels. The old Daly West shaft is an additional part of the ventilation system, as well as an emergency escape route.

There are three methods of ground support to shore the walls in the mine. Broken ground is covered with mesh wire and then secured with heavy bolts. Faulted areas that threaten with heavy ground are braced with wooden laggings or steel sets. Sand fill is used for mined-out stopes.

It used to be that miners took their lunch breaks astraddle a wooden lagging to keep them out of the water. Now, cave-like "dog houses" with benches, electric lights and heaters are provided.

Mining is never easy work, but modern equipment and safety measures have wrought many changes. In the late 1920's, drilling apparatus consisted of crank liners, "wiggle-tail buzzies" and jack hammers that had to be propped up, pushed in and rotated by sheer muscle while a cloud of rock dust threatened the miner with "Miner's Con." Now, they use a pneumatic, jackleg drill that is supported and pushed by a telescoping leg and powered by compressed air and a stream of water that eliminates the dust.

Before, in order to blast, they would ignite dynamite in twenty or thirty holes with the flame from the carbide light on their hats. Today, the series of rounds is connected by "spitter cords" so that it is only necessary to light one fuse.

Miners used to have to do all of the mucking and slushing of broken rock with a shovel. Now, such "goon spooning" is relegated to cleaning out ditches and other such maintenance work. Instead, an automatic mucking machine is operated by two levers which move a scoop bucket forward to fill up on a pile of ore and backward to dump the load into a storage bin or train car.

In the stopes, slushing machines retrieve the shattered ore after blasting. A large scoop-bucket is dragged back and fourth by lever-operated cables. Ore is scraped into a chute. From there, it is transferred into ore cars, trammed by rail to the hoist station and dumped into trenches and slushed into measuring pockets until there is enough to fill a container called a skip, which rides to the surface beneath the floor of the man cage.

Once the ore is topside, it is dumped onto a moving belt and carried toward the primary crushes at the mill. As the ore moves along the belt, an attendent removes excess debris and particularly fine specimens.

The primary, or jaw-crusher, is an immense V-shaped tub with one stationary side and one side that rocks back and forth to smash large chunks of rock and transfer them to another moving belt. From there, the ore travels to the secondary, or cone-crusher, where a cone-shaped apparatus spins unevenly inside of a large cylinder to further crumble the rock and drop it onto a conveyor belt en route to the storage bins.

Next, the gravel-sized rock is conveyed to the ball mill, a huge metal drum, where water is added and the ore is pounded into a fine mud by tumbling steel balls. Water and chemicals are then added to create a slurry state so that the liquid ore-carrying solution may be pumped into bubbling flotation tanks for extraction of the concentrates.

Once extracted, the concentrates are sent to thickener tanks where water and chemicals are removed and, from there, they are further dried on large filters. The final product is a greyish mud that turns to powder when it is dry. This is dumped in a semi-moist state into storage bins until removed by front-end loaders to be trucked to the railhead.

Shelley Christiansen working a pneumatic jackleg drill.

Miners still wear "hard-boiled hats" and yellow slickers.

A miner must be adept at repairing broken hoses and other equipment.

Shelley Christiansen works the levers of a slusher in her stope.

Very little is wasted. Water and sand that have been removed from the slurry are recycled by pumping the water back up the mountain to be reused in the milling process and utilizing the sand for mine fill. Other waste and overburden flows through a slime line for disposal in a tailings pond near Richardson Flat.

Mining and milling procedures have changed drastically over the years, but today's miner is little different from those of the past. He finds his work fascinating and enjoys the dark confines of his underground environment. He is a jack-of-all-trades, called upon to act as plumber, electrician, mechanic, carpenter or track-layer, according to what is needed at his working place. He develops fierce loyalties to others in his trade. In fact, on September 11, 1975, the *Park Record* reported a wildcat strike at the Ontario the week before. One of the miners had had a run-in with his foreman and was fired. The miner claimed that the problem was merely a personality conflict and the management finally decided to retain him but gave him a three day suspension. His buddies claimed this was unfair. They were so incensed by the incident that they walked out en masse and closed down the entire operation for three days in reprisal.

But after the hard work is done, when dripping yellow slickers are removed and grime is showered away, the miners head for Main Street and The Cozy for a schooner of brew, a game of pool and a verbal joust of mining stories.

Meanwhile, up on the hill at the head office, the "big boys" are analyzing geologic maps and cores of drilled rock. Here, there is some high-grade mineralization with a good promise of ore; there, a potential for deep copper; over on the west, perhaps silver. They know it's down there and they're just as determined to find it as the roaming prospectors of yore. Park City's diggings and doings are bound to continue for many years to come.

Miners enjoy their free time in exciting Demolition Derbies.

Portfolio

by Harry Harpster

A Photographer's Response to the Park City of Today

"She's a good little town in a good little holler."

SILVER KING
COALITION MINES
CO.

SALT LAKE HOUSE

PARK AVE
MAIN ST

THE
PROSPECTORS
HOTEL
SALOON

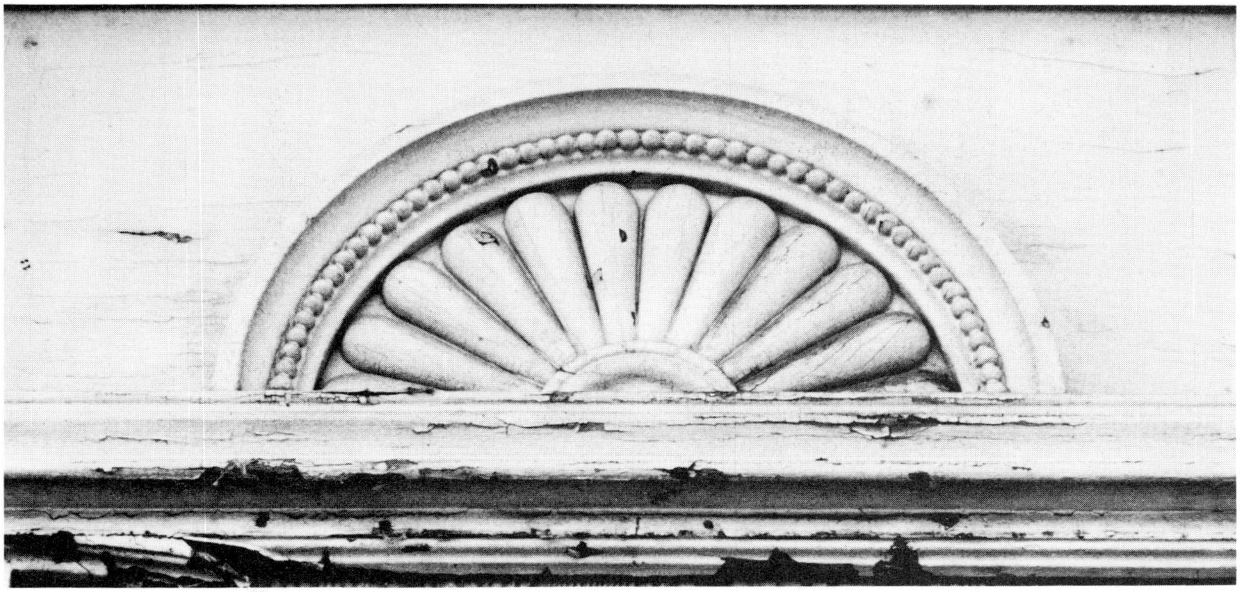

St. Luke's
Episcopal
Church

539

POLICE
528

CITY HALL
1885

57

VICTORY
Chlorodyne

DAN MCPOLIN
PARK CITY
BOTTLING WORKS
UTAH

DANNY

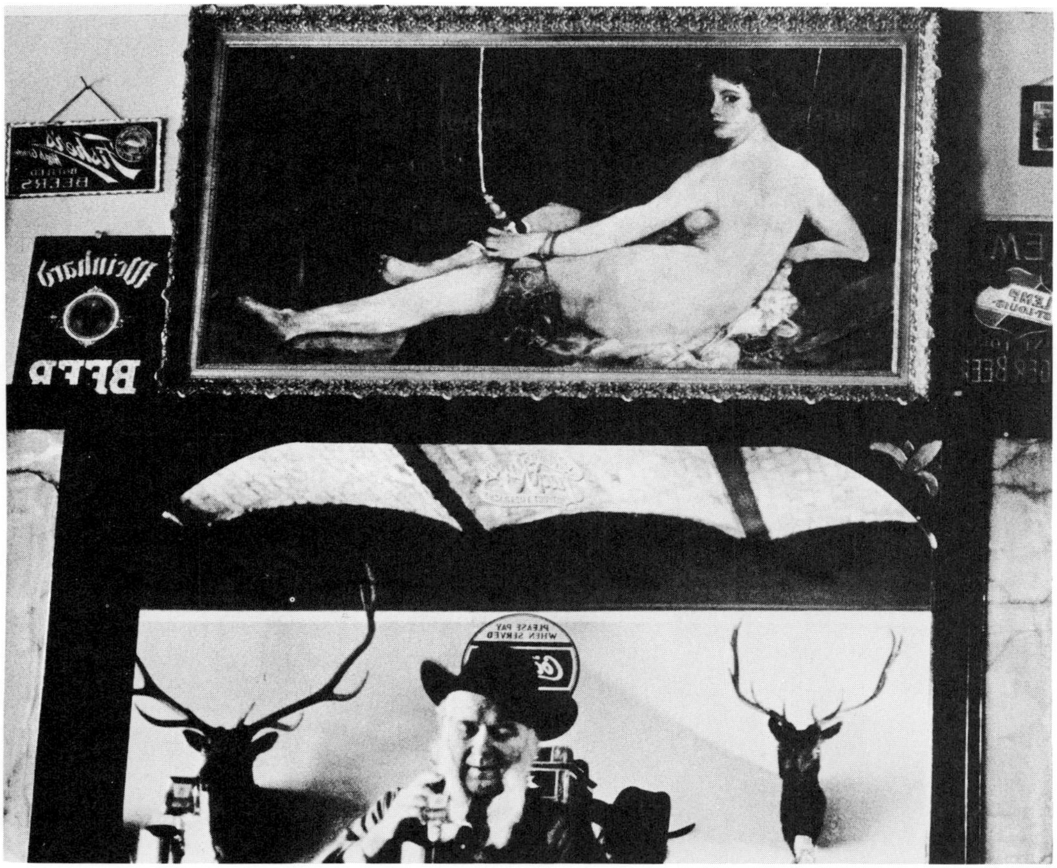

POISON CREEK
RESTAURANT

UNION PACIFIC
PARK CITY
NORTH
224
RAILWAY EXPRESS AGENCY

THE
WATERBED CO.
434

SILVER
WHEEL
NOW PLAYING
FRI MELODRAMA SAT
LITTLE NELL

MAIN FURNITURE & HDW

B S
BARBER SHOP

The
BUCKET

PARK CITY
BALBOA
LOS ANGELES
POISON CREEK
RESTAURANT
OAK

Creative
Glassblowing
by Bruno

359

THE
OWL'S
ROOST

Park City... Recipes

As in its heyday, contemporary Park City boasts good cooks of many nationalities. A number of them have shared traditional favorites by contributing the recipes in this section. The recipes are divided according to their places of origin, and are followed by some that have no nationality, but are JUST PLAIN GOOD.

English

CORNISH (COUSIN JACK) PASTY

This recipe for Cornish pasty was brought from the Isle of Man many years ago. Like the old Cornish miners who carried a pasty for noon lunch, early Park City "Cousin Jacks" also carried pasties in their "pie-cans."

MEAT FILLING

1 lb. veal or round steak, cut in ⅜-inch cubes
1 lb. pork steak, cut in ⅜-inch cubes
2 potatoes, diced
2 carrots, sliced
2 T. chopped onion
Pepper and salt
Butter

PASTRY

2 cups flour
¾ cup shortening
½ tsp. salt
Cold water

Roll pastry, using 8-inch plate as a guide. Cut out 5 even rounds. On each round arrange a layer of uncooked potatoes, a layer of uncooked meat, a layer of carrots and onions. Season each layer with salt and pepper. Dot with butter. Fold half of pastry circle over the mound. Press edges together, piercing the top to let out steam. The pastry will resemble a turnover. Place in baking pan and bake 20 minutes at 400°, then 45 minutes at 325°.

— *Mrs. Elden Sorensen*

CORNISH PASTY

MEAT FILLING

2 slices side pork or pork chops, ground
1 lb. round steak with a little fat, ground
1 large onion, chopped
1 large potato, cut fine or coarsely ground
Salt, pepper, poultry seasoning, celery salt to taste

HOT WATER PASTRY

1½ cups flour
½ tsp. salt
¼ cup boiling water
½ cup shortening (scant)
½ tsp. baking powder

Heat bowl with hot water. Place shortening in bowl and stir till smooth. Sift salt, baking powder, and flour together and add to shortening. Stir until a smooth, round ball forms on spoon. Chill before rolling out (if chilled, dough keeps well for several days). Roll pastry very thin. Cut into 6-inch rectangles. Put layers of meats, onions, potatoes, and seasonings in the center of each rectangle. Dot with butter. Fold both sides over mound; pinch sides together, leaving opening in top. Bake 30–35 minutes at 450°. If there is not enough juice to keep pasties from sticking to pan, add small amount of water when they are half cooked.

— *Mrs. Josie Hays*

SAFFRON BREAD

1 gram Spanish saffron
1 cup boiling water
2 cakes yeast
2 cups milk, scalded
6 cups sifted flour
1 tsp. salt
1 cup sugar
½ cup melted shortening
1 cup currants
1 cup chopped lemon peel (or citrus fruit mix)

Place the saffron in the boiling water; let steep until cool. Dissolve yeast (according to directions on package). Scald the milk and let it cool. Mix the flour, salt, and sugar, then add melted shortening and mix well. Knead until smooth and elastic (a little more flour may be needed). Add the currants and lemon peel. Allow to rise until double in size; pan, and let rise again (about 30 minutes). Bake 40–45 minutes at 375°.

— *Mrs. Harold Williams*

BROWN BREAD

1 cup dates
1 cup warm water
1 tsp. soda
1 cup brown sugar
2 T. shortening
¼ tsp. salt
2¼ cups sifted flour
2 eggs
1 tsp. vanilla
1 cup coarsely chopped black walnuts
1 jigger rum

Cut dates into small pieces and soak in water with soda. Cream together brown sugar and shortening. Add the salt, flour, eggs, vanilla, walnuts, and rum. Add dates and bake 45–60 minutes at 350°.

— *Mrs. Josie Hays*

ROAST BEEF WITH YORKSHIRE PUDDING

3 oz. sifted flour
¼ tsp. salt
2 eggs, beaten
About ½ pt. milk
2 T. roast beef drippings

Approximately 1 hour before a roast of beef is done, sift flour and salt together. Make a well in center of flour; add beaten eggs, then add sufficient milk to make a thin batter. Beat well. Place drippings in a pan or pie dish; heat until smoking hot and pour in the batter. Bake about 1 hour at 400°. Serves four; increase ingredients for additional servings.

— *Mrs. Fred Eley*

BEEFSTEAK AND KIDNEY PIE

½ recipe of pastry
2 lbs. steak
1 or 2 beef kidneys
1 onion
1 T. flour
Salt and pepper
About ¾ pt. water

Remove fat from steak and skin from kidneys. Cut kidneys into small pieces, and meat into 1-inch squares. Roll in seasoned flour. Melt about 1 oz. of drippings in a saucepan and brown meat and sliced onion. Drain off any fat; add water and simmer 1½ hours. Leave until cold. Put into pie dish, adding only a little of the gravy. (Keep the remainder to serve with pie.) Cover with pastry, leaving a hole in the center. Bake 40–45 minutes at 425°. Serves six.

— *Mrs. Fred Eley*

CHICKEN AND DUMPLINGS

2 cups flour
½ tsp. salt
1 heaping tsp. baking powder
2 eggs
¾ cup milk (or enough to make a rather dry mixture)
1 T. melted butter (less if the broth is very rich)

Stew chickens according to favorite recipe. Add salt and baking powder to sifted flour. Beat eggs slightly with a fork and then add to milk; add to flour mixture gradually. When half the milk is added, stir in the butter and add remaining milk. (Cook in very large pot, as batter will boil over if it doesn't have room to swell.) Drop by spoonfuls into rapidly boiling broth. There must be about 4 cups of broth as dumplings absorb a great deal. Cover with tight-fitting lid and never peek, because they will be soggy if you remove the cover before they are cooked. If you drop batter by tablespoons, cook for 15 minutes; if you make dumplings smaller, cook 12 minutes. Serve at once, with broth if desired. Serves four, with seconds (which they will want).

— Lucille Iredale Carleson

SINGING HINNY

An old family recipe brought from Northumberland, England — especially for birthdays.

4 cups flour, sifted
2 tsp. baking powder
1 tsp. salt
4 T. sugar
1 cup shortening
1 box currants (¾ lb.)

Mix ingredients and moisten with cold water or milk. Roll about ½-inch thick. Mark in squares and bake 20 minutes at 400°. Split in two and butter. Serve hot with currant jelly.

— Lucille Iredale Carleson

SUET PUDDING

1 cup ground suet
1 cup hot water
½ cup nuts
1 cup brown sugar
1 cup raisins
2 cups flour, sifted
¼ tsp. salt
2 tsp. baking powder
1 tsp. each of cinnamon, cloves, and nutmeg

Mix ingredients together and put in pudding molds. Place molds on a rack in a heavy, covered pan with about an inch of water. Steam 2 or 3 hours. Serve with rum sauce or hard sauce.

— Mrs. Charles Himes

Irish

BEEF STEW

1 lb. beef cubes
1 T. oil
1 large onion
6 large potatoes, cut in quarters
6 medium carrots
4 large stalks of celery
1 can peas
3 T. cornstarch

Sauté beef cubes in fat until brown. Add water to cover and boil 2 hours. Add carrots, onions, and celery, all cut in large pieces. Cook potatoes in separate pan. When done, add to meat and vegetable mixture. Thicken with cornstarch. Add peas; salt and pepper to taste. There will be enough liquid to use as a gravy.

—Jesse McAlevy

CLAM CHOWDER

1 can Pioneer Sea clams
3 medium potatoes, diced
1 medium onion, diced
1 can cream of celery soup
4 slices bacon
1 cup milk

Cook bacon, then add onion and brown in drippings. Drain fat from mixture. Add cream of celery soup and milk, potatoes and clams. Simmer until potatoes are tender. Season to taste with salt and pepper.

—Jesse McAlevy

Greek

STEFATHO

2 lbs. boneless chuck beef
4 lbs. small onions
½ cup olive oil
1 tsp. whole pickling spice
1 can tomato paste
5 cloves garlic
Salt and pepper to taste

Cut meat in pieces; brown slightly in oil. Peel onions and leave whole. Add to meat together with spices (in a metal tea ball or tied in a clean cheesecloth), garlic, tomato sauce, one can of water and seasonings. Cover tightly and cook until meat and onions are tender and liquid is reduced to a delicious gravy. Remove spices at once.

—Catherine Polychronis Goff

BEEF OR LAMB KABOBS

Use leg of lamb or have butcher cut a choice or prime round steak 1¼-inches thick. May use any other cuts of beef not less tender. Remove fat and any large sinews or gristle. Cut meat into 1¼-inch cubes. Slice fat into wafers to intersperse next to meat cubes on skewers. Place meat and fat in amply large bowl or pan. For about 4 lbs. meat, sprinkle with:

4 tsp. oregano
4 T. olive oil
4 tsp. salt
2 tsp. pepper
2 tsp. garlic granules or comparable amount crushed garlic, garlic powder, or garlic salt. If using garlic salt, reduce salt to 2 tsp.

Juice of 1 very juicy large lemon or 2 regular lemons. Increase lemon juice by half if using lamb.

Knead into meat by hand until spices and liquids are thoroughly distributed. Let stand at room temperature a couple of hours for flavors to penetrate. Mix occasionally to distribute any liquids settling to bottom. Thread cubes and fat slices on skewers, interspersing with good-sized raw mushrooms, green pepper slices, cherry tomatoes, chunks of raw onion, or 1-inch pieces of raw corn. (For 4 lbs. meat, 1 lb. mushrooms, 3 green peppers sectioned in sixths, and 1 lb. cherry tomatoes or 5 medium tomatoes, quartered. Do your own thing.) Broil over grey charcoal or under broiler to suit individual preference. Serves 8.

— *Mary Lehmer*

SPANAKORIZO

1 lb. spinach
2 medium chopped onions
1 can tomato sauce

¼ cup rice (not instant)
¼ cup olive oil
Salt and pepper to taste

Sauté the onions in heated oil until golden. Add the spinach which has been thoroughly washed and drained. Add tomato sauce and 1 can water; bring to boil; add rice. Stir, season to taste, and simmer about 15 minutes or until tender. Serve with a lemon slice or juice, added to taste.

— *Catherine Polychronis Goff*

Scandinavian

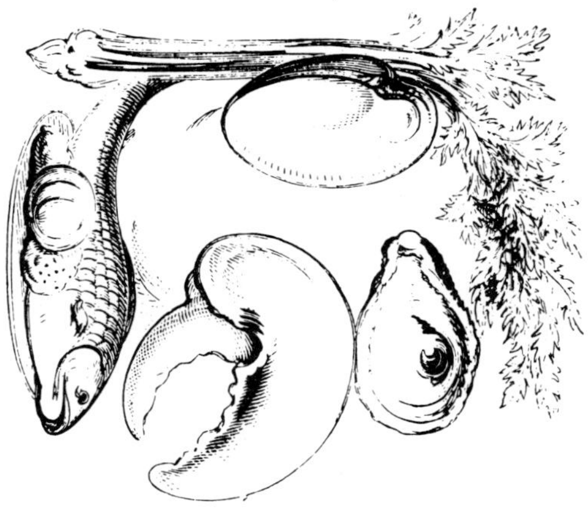

FISK PANNA

4 potatoes, cooked and sliced
1 onion, sliced
20 oz. kippered salmon
Milk
Salt and pepper

In casserole, place a layer of sliced potatoes, a layer of onion, a layer of kippered salmon, and another layer of potatoes. Season with small amount of salt and pepper. Add milk to cover. Dot with butter. Cook at 350° about 1 hour, or until the butter browns on top and bubbles through.

— *Mrs. Nina Lund*

FISH CHOWDER

3 or 4 potatoes, cut in pieces
1 onion, chopped
Salmon or trout, boned and cut in pieces
½-cup canned milk
6–7 whole allspice
Butter
Flour

Boil potatoes and onions in small amount of salted water until almost done. Put fish pieces on top of potatoes. Dot with butter. Add milk and allspice, and finish cooking (about 15 minutes). Thicken with flour and water.

— *Mrs. Eric Joffs*

LIMPA BREAD

1 qt. milk
1 T. salt
8 T. sugar
3 yeast cakes
8 T. shortening
3 T. anise seed
3 T. fennel seed
8 T. rye flour
Enough molasses to produce a desirable color
Enough white flour to make a medium-stiff dough

Mix as white bread, let rise to double in bulk. Make into round loaves, place in 9-inch cake pans, and pierce dough with fork. Let rise 1 hour. Bake at 300° for 1 hour. Remove from pans and brush tops and sides of loaves with glaze made of sugar and water which have been boiled together until sugar is dissolved. Makes 4 loaves.

— *Mrs. Nina Lund*

STOCKFISH (LUTE FISK)

Traditional Christmas Eve dinner

Dried Norwegian Stockfish is as hard as kindling. Buy a whole fish and with a saw or cleaver split and cut into pieces. In a wooden bucket or galvanized wash tub, soak fish in plain cold water for 10 days, changing water at least every other day. Then fix a lye bath of 2 gal. warm water and ½ cup lye (or strong enough solution to burn your tongue). When lye has dissolved and cooled in water, soak fish for 48 hours (or until you can put your finger through to the back). A bath of 2 gal. water, 2 cups unslacked lime, and 1 small package sal soda may be used instead of the lye bath, if desired. Fish will come out thick and fluffy. Soak again in plain cold water for 2 or 3 days, changing water every day. Fish is now ready to cook. (If you can't find dried stockfish or don't want to spend the time soaking it, the fish may be purchased ready for cooking). Put fish in boiling, salted water to cover. Add a few kernels of whole allspice. Boil 5 or 10 minutes. Cover with thick white sauce and sprinkle with powdered allspice, or use mustard sauce. Serve over boiled potatoes.

—*Mrs. Eric Joffs*

CORN SALAD

1 can yellow cream-style corn
2 T. mayonnaise
2 or 3 chopped green onions
Chopped iceberg lettuce
2 hard-boiled eggs, sliced
Dash salt, pepper, and paprika

Mix ingredients together and garnish with sliced eggs and paprika.

—*Mrs. Nina Lund*

SWEDISH FRUIT SOUP

1 cup each dried prunes, apricots, apples, peaches, raisins
1 cup rice
2 T. maple syrup
3 T. sugar
1 stick cinnamon
Pinch of salt
Cornstarch
1 T. vinegar

Soak dried fruit in 4 quarts cold water overnight. Simmer with rice about 1 hour, then add remaining ingredients and thicken with cornstarch. Serve warm.

—*Mrs. Eric Joffs*

SWEDISH OVEN PANCAKE

2 or 3 T. butter or margarine
1 pt. milk
9 eggs
6 T. sugar
6 T. flour
1 tsp. salt
Dash cinnamon

Melt butter in pan, turning pan so that the melted butter goes all around sides and bottom. Place a pan of water in oven. Mix all ingredients and beat until well blended. Pour in buttered pan and bake 45 minutes at 400° ; or until knife inserted in center comes out clean. When slightly cooled, sprinkle with sugar and cinnamon. The flour will settle to the bottom and the butter will form a crust on top. The mixture rises high during baking and falls as it cooks. A round roasting pan 4 inches high and 10 inches in diameter is recommended for cooking.

— *Mrs. Nina Lund*

Italian

MRS. PEDROTTI'S MINESTRONE

2 cups Italian dried beans, or pinto beans
1 small ham hock
2 stalks celery, cut in ½-inch pieces
2 carrots, diced
2 medium-sized potatoes, quartered
1 medium-sized zucchini, cut in pieces (or a piece of squash)
¼ cup butter
1 medium onion, chopped
2 T. parsley, chopped
½ cup celery leaves, chopped
1 tsp. salt
¼ tsp. pepper
½ cup tomato sauce
1 cup small shell or other small-cut macaroni

Wash beans thoroughly, discarding imperfect beans. Soak for 3 or 4 hours in cold water, then add enough water to make 5 qts. in a kettle of at least 6 qt. capacity. Bring to a rapid boil, then turn heat to simmer for at least 1 hour. Add ham hock and continue cooking 1–1½ hours. Add celery, carrots, potatoes, zucchini; let soup come to a boil and continue cooking slowly. Meanwhile, sauté in butter the onion, parsley, celery leaves, salt and pepper. When onion is golden color, add the tomato sauce and continue cooking slowly about 15 minutes. Add this to soup and simmer. About ½ hour before beans are cooked, add macaroni.

— *Mrs. John Crus*

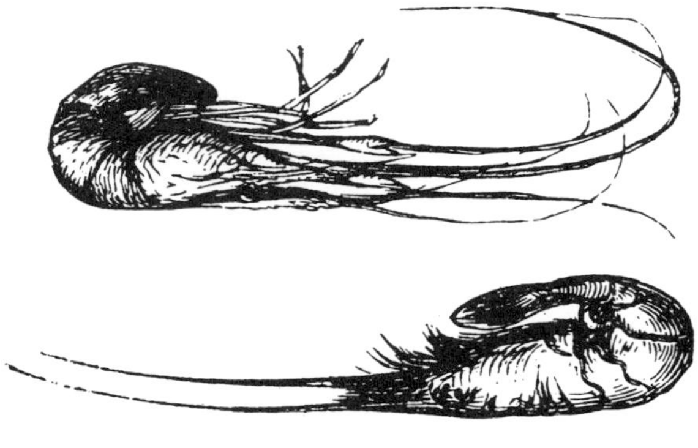

VEAL STEAK IN CASSEROLE

1 veal round steak 2 inches thick
white wine
1 small onion, chopped
¼ lb. prosciutto (cut in small pieces) or ham
salt and pepper
2 T. butter
½ cup dried mushrooms (soaked in warm water about ½ hour)
1 cup broth or water

Marinate veal in enough dry, white wine (to which salt and pepper have been added) to cover, for about 1 hour. Then, in deep skillet, sauté onion and proscuitto in butter until onion is tender but not brown. Add meat and brown lightly. Add mushrooms and wine marinade. Let wine evaporate, then add 1 cup broth or water, cover and cook slowly until meat is tender (½ to ¾ hour).

— *Mrs. John Crus*

Chinese

FRIED CHICKEN — CANTONESE STYLE

1 whole frying chicken
Salt
Flour
1 stalk of green onion
1 clove garlic
1 tiny piece of fresh ginger
1 tsp. salt
4 T. soy sauce
1 tsp. wine, sherry, or liquor
1 cup water

Cut chicken into medium-sized pieces, dip in seasoned flour, and fry to light, golden brown. Cut the onion and garlic up fine; mix with the ginger, salt, soy sauce, wine, and water. Add the mixture to cooked chicken. Bring to a boil and simmer 10 minutes, stirring occasionally. Serve with plain steamed rice and a leafy vegetable.

— *Oay Gin Chinn*

FRIED PRAWNS

1 lb. shrimp
½ cup flour
2 T. baking powder
½ tsp. salt
2 tsp. soy sauce

Shell shrimp, cut in half lengthwise, and clean. Mix flour, baking powder, salt, and soy sauce together, then dip shrimp in batter. Deep fry until golden brown.

— *Oay Gin Chinn*

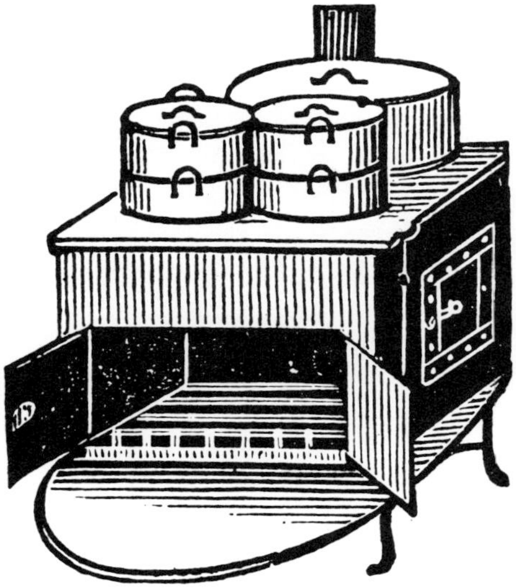

GREEN PEPPER TOMATO CHOW YUK

½ lb. beef, sliced
1 green pepper, chopped
½ onion, chopped
1 whole tomato, quartered
2 T. catsup
1 tsp. salt
1 tsp. sugar
½ cup water
Cornstarch

Brown beef. Add green pepper, onion, catsup, water, salt and sugar, and bring to boil. Thicken with cornstarch, then add tomato, bring to boil, and serve.

— *Jim Goo, Sr.*

ALMOND COOKIES

1 cup lard
2 cups sugar
1 egg
1 tsp. almond flavoring
1 tsp. lemon flavoring
3 cups flour

Mix ingredients well. Roll into small balls. Cook over water in oven at 300° for 15 minutes.

— *Jim Goo, Sr.*

Just Plain Good

WEDDING RING SALAD

1 cup mandarin oranges, diced
1 cup pineapple, diced
1 cup bananas, diced
1 cup coconut
1 cup miniature mashmallows
1 cup sour cream

Mix together fruits, coconut, and marshmallows and fold into sour cream. Chill and serve on lettuce leaf.

— *Jesse McAlevy*

SPANISH GREEN BEANS

1 onion, minced
1 green pepper, chopped
1½ T. sugar
1 can tomato soup
1 can green beans
4 strips bacon, fried crisp

Brown onion and green pepper in bacon fat. Add sugar and soup. Mix with beans and bacon. Put in greased casserole and bake at 350° for 1 hour. Serves six.

— *Mrs. Charles Himes*

CORN, MACARONI AND CHEESE

Boil macaroni in salted water. Drain. In greased casserole, place a layer of macaroni, layer of sliced mild cheese, layer of yellow creamed corn, salt and pepper, another layer of macaroni, and top with cheese cubes and buttered crumbs. Pour milk over, let seep through. Bake about ½ hour at 350°.

— *Mrs. Nina Lund*

BEEF A LA KING

1 lb. ground chuck
Green onion, chopped
1 can undiluted mushroom soup
1 can undiluted cream of chicken soup
½ pint sour cream
Rice

Brown meat with onion, add soups, simmer until ready to serve, then add sour cream. Serve over rice. (Meat and soups may be prepared two or three hours in advance. Cook rice and add sour cream to meat mixture just before serving.)

— *Olive Blackler Potter*

MOTHER'S POTTED MEAT

2 lbs. beef heel
2 lbs. lean pork shoulder
1 tsp. salt
½ tsp. sage
¼ tsp. marjoram
¼ tsp. thyme

Place meat in large pan. Cover with water and add salt. Let simmer until meat leaves bones. Remove meat and place in large mixing bowl. Reserve liquid. With two sharp knives, chop meat very fine. Add sage, marjoram, and thyme. Add enough meat gravy to make a not too thin mixture (about 2 cups). Taste for seasoning. Place mixture in a 3 lb. coffee can or other container. Let stand in refrigerator several hours or until firm. Turn mold on plate upside down and tap lightly to loosen meat. Slice in round, thin slices. Keep refrigerated.

— *Clara Shea*

SOLAR TEA

Fill a 1 gallon glass container with cold water and add 7 tea bags. Cover and set it in the sun to steep to desired strength (approximately 4–8 hours). For flavor, add ¾ cup sugar and lemon juice to taste. Keep refrigerated and use in 2 or 3 days.

— *Dorie Spurlock*

GOOD VEGETABLE SOUP

3 or 4 lbs. soupbone or beef ribs
1 cup pearl barley
1 tsp. salt
⅛ tsp. black pepper
6 or 8 carrots chopped small
2 or 3 turnips, chopped
1 large or 2 small onions, chopped
Small bunch celery, chopped
1 package frozen peas
1 lg. can tomatoes
¼ head cabbage, shredded

Place meat in large pan and cover with water. Simmer 3 or 4 hours. Add remaining ingredients and simmer until vegetables are done.

— Clara Shea

SOFT MOLASSES COOKIES

2½ cups flour
1 tsp. ginger
1 tsp. cinnamon
½ tsp. salt
2 tsp. baking soda
½ cup shortening
½ cup sugar
½ cup light molasses
1 egg
Raisins (optional)

Sift together flour, ginger, cinnamon, and salt. Dissolve baking soda in 2 T. hot water. Blend shortening, sugar, molasses, baking soda, and egg and mix until light and fluffy. Beat in flour and spices, alternately, with 6 T. cold water. Drop by teaspoons on greased cookie sheet, 2 inches apart, and place 4 or 5 raisins on top of each, if desired. Bake 8 to 10 minutes in 400° oven. Remove with wide spatula and cool.

— Clara Shea

WACKY CAKE

1 cup sugar
1½ cups flour
½ tsp. salt
1¼ cups cocoa
1 tsp. baking soda
1 T. vinegar
1 tsp. vanilla
⅓ cup salad oil
1 cup cold water

In an ungreased square or oblong pan (9 by 13 by 2 inches is a good size), sift sugar, flour, salt, cocoa, and soda. Then add vinegar, vanilla, salad oil, and water. Now all you have to do is stir this mixture around until there are no lumps. Bake 30 or 35 minutes at 350°.

— Eleanor G. Bennett

SAVORY SEASONING SALT

1 lb. salt
3 T. pepper
3 T. paprika
9 T. MSG

Mix by shaking and sprinkle on foods liberally before cooking.

— *Steve Stevens*

YUM-YUM FRUIT CAKE

In a large saucepan, combine:

1 lb. seeded raisins
1 lb. seedless raisins
2¾ cups sugar
2 cups water

Bring to a boil. Boil just 1 minute; take from fire. Add:

¼ lb. butter
¼ cup shortening
½ package candied fruit mix

Let this mixture cool. Add:

1½ cups cold water
3 tsp. cinnamon
2 tsp. nutmeg
1 tsp. mace
1½ tsp. soda
½ lb. shelled walnuts
2 unbeaten eggs
4½ level cups unsifted all-purpose flour

Blend well and carefully. Grease 4 loaf pans; sprinkle with flour (not necessary to line pans with paper). Bake 1 hour at 350°, then turn off oven and leave pans in 15 minutes longer. Remove from oven, but leave cakes in pans 15 minutes more before removing.

— *Mrs. Nina Lund*

MARSHMALLOWS

2 T. unflavored gelatin
2 cups white sugar
¾ cup water
2 eggs
1 tsp. vanilla

Soften gelatin in ½ cup cold water and place in large bowl. Make syrup of the sugar and water, and boil to soft ball stage. Beat 2 egg whites stiff in a smaller bowl. Put beater in bowl with gelatin and then pour syrup over slowly while beating. Beat about 5 minutes or until stiff. Add egg whites and beat; then add vanilla (and food coloring, if desired). Pour mixture into ungreased square pan. Let set overnight, then cut with a knife dipped in hot water. Marshmallows can be rolled in toasted nuts, coconut, etc.

— *Mrs. Charles Himes*

DEER MINCEMEAT

8 cups lean cooked deer meat, ground
16 cups ground apples
2 cups ground suet
3 packages raisins
1 package candied fruit
3 packages currants
2 tsp. salt
1 T. cloves
1 tsp. mace
1 qt. grape juice
2 lbs. brown sugar
1 cup white sugar
1 qt. cider or 1 cup vinegar
1 T. each cinnamon, nutmeg, and allspice

Mix well and cook until apples are tender. Meat stock may be added. Put in jars, place in pressure canner for 20 minutes at 10 lbs. pressure or a water bath for about 1½ hours. Good in cookies, pies, etc.

—*Mrs. Charles Himes*

FRUIT COCKTAIL SQUARES

1 cup shortening
3 eggs, beaten
1 cup raisins
1 cup nuts
1 tsp. soda in 3 T. hot water
1½ cups sugar
3 T. molasses
1 cup fruit cocktail
2 tsp. vanilla
1 tsp. cinnamon
½ tsp. nutmeg
1 tsp. salt
1 tsp. cream of tartar
3½ cups flour

Mix as listed. Bake at 350° about 20 minutes. Frost and cut into squares.

—*Mrs. Fred Eley*

DUTCH APPLE PIE

½ recipe pastry
1 lb. (No. 303 can) pie apples
½ can water
1 cup sugar
¼ cup cornstarch
¼ tsp. salt
½ tsp. cinnamon
½ tsp. nutmeg
Juice of ½ lemon

Bring apples and water to boil and add dry mixture. Boil 5 minutes and pour into uncooked pie shell.

TOPPING

1 cup brown sugar
¼ cup flour
¼ tsp. salt
¼ cup butter or margarine

Blend well and cover top of apple mixture. Bake at 400° for 30 minutes, or until browned.

—*Jesse McAlevy*

SALAD DRESSING CAKE

1 cup dates
1 cup nuts
1 tsp. soda
1 cup hot water
¾ cup salad dressing
1 cup sugar
2 T. cocoa
1 tsp. cinnamon
24 maraschino cherries
¼ tsp. salt
1 tsp. vanilla
2½ cups sifted flour

Pour hot water over dates, nuts, and soda. Let stand till cool. Add salad dressing, sugar, cocoa, cinnamon, cherries, salt, vanilla, and flour; mix. Bake in 9 by 13 inch pan at 350°, for 45 minutes.

—Mrs. Charles Himes

ELDERBERRY JELLY

Pick, wash, and stem berries (they grow all around Park City in August and September). Cover with water and boil. After they come to a boil, set pan aside for 10 minutes. Mash with ricer or potato masher, and strain twice through cheesecloth. Combine:

4 cups elderberry juice
¼ cup lemon juice
5 cups sugar
1 package pectin

Bring to a fast boil and boil exactly 2 minutes. Remove from heat, skim, and fill jars. Cover with paraffin.

—Mary E .Grose

CORNCOB JELLY

Corn you buy in the store is not suitable for this. You must use corn which has dried on the stalks; shuck it, and remove kernels.

12 corncobs, washed
5¼ cups corncob juice
7 cups sugar
1 package pectin
Juice of 2 lemons
(Vegetable coloring, optional)

Break cobs into pieces, cover with water and boil until water is coffee brown. Strain juice in cheesecloth jelly bag. Measure 5¼ cups of juice and add other ingredients. Boil 2 minutes, remove from heat, skim, and bottle.

—Kendall Webb

LEMON MERINGUE PIE

CRUST

3 cups flour
¼ tsp. salt
1 cup lard
½ cup water

Blend flour, salt, and lard. Add water. Roll out and line pie plate. Bake at 400° for 15 minutes.

FILLING

2 cups water
1 cup sugar
¼ cup cornstarch
Juice of 2 lemons
¼ tsp. salt
1 T. butter

Bring water to a boil. Mix sugar, salt, cornstarch, and lemon juice and add to boiling water. Boil until mixture thickens. Add butter and pour into baked pie shell.

MERINGUE

3 egg whites
½ tsp. cream of tartar
¼ cup sugar

Beat egg whites and cream of tartar until stiff. Add sugar and beat until whites stand in peaks. Cover pie with meringue and bake at 400° until brown.

— *Jesse McAlevy*

RAISIN FILLED SQUARES

DOUGH

½ cup butter or shortening
1 cup sugar
1 egg, well beaten
1 tsp. vanilla
½ cup milk
3 cups flour
1 tsp. salt
3 tsp. baking powder

Cream shortening, add sugar gradually then add well-beaten egg and vanilla. Sift in dry ingredients and add milk slowly to make a soft dough.

FILLING

¾ cup sugar
1 cup raisins, chopped
1 cup boiling water
2 T. flour

Cook above ingredients until thick. Roll half the dough to ¼ inch thickness and place in bottom of a greased baking pan. Spread filling to ½ inch thickness and cover with remaining dough rolled to ¼ inch thickness. Bake 45 minutes at 375° or until dough is firm to touch. When partly cooled, cut into squares.

— *Mrs. Nina Lund*

HONEY PIE

1 9-inch pie shell and top crust
1 cup raisins
1 cup cranberries
1 cup sugar
1 cup cream
Butter

Mix ingredients and place in pie crust. Dot with butter and put on top crust. Bake at 400° for approximately 40 minutes.

—Jennifer Giddings

ICEBERG SLICES

Also for green tomatoes

8 lbs. cucumbers
2 gallons water
2 cups lime
2 qts. vinegar
9 cups sugar
2 T. salt
1 tsp. celery seed
1 tsp. whole cloves
1 tsp. allspice

Wash and slice cucumbers. Soak 24 hours in water with lime added. Put in crock. Bring to boil vinegar, sugar, salt, celery seed, cloves, and allspice. Add cucumbers (which have been washed 2 or 3 times) to hot syrup. Let stand 3 hours, then boil ½ hour and can.

—Mrs. Charles Himes

ROSLYN'S POTPOURRI

½ oz. violet-scented talcum powder
1 oz. orrisroot
½ tsp. mace
½ tsp. cinnamon
½ tsp. cloves
4 drops oil of rose geranium
2 sprigs dried eucalyptus leaves
2 tsp. finely grated dried orange peel
2 qts. dried rose or carnation petals

Mix first 6 ingredients in a bowl. Very slowly add the eucalyptus leaves and the orange peel. Sprinkle the mixture between 2-inch layers of the dried flower petals in a glass jar and cover tightly. Tie in net bags to give a wonderful fragrance to linen closets, drawers, and chests.

—Roslyn Grose

ELDERBERRY WINE

A delightful and refreshing homemade wine that was very popular with old Parkites.

Pick 3 lbs. of ripe elderberries. Do not wash them, as you will wash off the natural fruit yeast. Put berries in small crock large enough to hold berries and 1 gallon water. Smash berries, add the gallon of water and 1 lb. sugar. Cover with clean cloth and let stand for 24 hours or until the fermentation has started strongly. (The proper temperature for making wine is between 65 and 75°.) Now gather elderberries for the main batch of wine. Weigh the berries after removing half the stems (the tannic acid content will be too high and the wine too bitter if they are left on). Put the berries in a large crock and cover with water. Boil berries gently for ½ hour. Strain off the liquid through a strainer and press out the pulp by placing it in a clean cloth and wringing it out. Measure the liquid and add enough water (spring water preferably) that the liquid averages 3 lbs. of original fruit to each gallon of liquid. Place in fermenting crock; add 18 oz. of sugar for each gallon of liquid. (This will make approximately 12 percent alcohol wine.) Now add the fermenting batch to the crock. Cover crock with wooden cover that has 10 or 12 holes drilled in it and cover with a clean dish cloth. Allow mixture to ferment for 3 or 4 days. Siphon liquid into clean gallon jugs. Wipe tops of jugs with clean cloth and plug with a wad of absorbent cotton. This wine is still fermenting. Let it stand for a month. If the jugs don't stay full, fill them with boiled water that has cooled.

After 1 month, siphon wine into another clean jug, being careful not to disturb the sediment at the bottom of the first jug while siphoning (let the wine run against the side of the bottle to aerate it). Cork the jugs tightly and let stand in a very cool place for another month. (Freezing will not hurt the wine now, but will help it.) Resiphon the wine into bottles you wish to serve it in. Be careful not to disturb the sediment and do not aerate the wine this time. Fill bottles to within ½ inch of the top and seal very tightly. The wine will not be ready to drink for another month and will be best after 1 year. If care is taken, you will be able to enjoy a different and flavorful wine. (There is always the possibility that your friends will enjoy a bottle of different and flavorful vinegar!)

—Edwin Grose

SOAP FLAKES

4 lbs. or 2 qts. melted fat
3 qts. less 1 cup cold water
1 can lye
1 cup commercial bleach
1 cup borax

Mix all ingredients together and stir occasionally until almost dry. Mixture will flake and be crumbly. Can be stored in empty milk cartons.

—Mrs. Charles Himes

LAUNDRY SOAP (BAR)

Put 1 can lye in 2½ pts. of water in an enamel pan. Stir until dissolved and allow to cool. Put 6 lbs. grease in double amount of water. Bring to a boil and boil for a few minutes to take out the salt (if grease contains no salt, boiling is unnecessary). Drain off water; let grease melt, then cool. Add lye and water slowly to the grease and stir thoroughly for 15 minutes. Pour into an enamel or wooden box. Keep in warm place for 2 days, then cut into cakes.

—Mrs. Charles Himes

Acknowledgments

As *Diggings and Doings in Park City* goes into a third edition, I want to reiterate my thanks to all of those who contributed so greatly to my research and collection of historic photographs — James Ivers, Sr. and Oscar N. Freindly (now deceased), Miss Hollister Hancock, David Keith, Susanna Hartman, Colleen Kearns Steiner, Dr. Harold Lamb, Judge Willis A. Ritter and James Ivers, Jr.

My appreciation also extends to those recent Parkites who helped so much, — Rick Hurst and Bill Norem of Park City Ventures, Sheldon Horrocks, Jack Johnson, Warren King and Max Jarman.

A special thanks to Mr. Lowell Thomas for writing another introduction, to Harry Harpster for his beautiful Portfolio, and to K. C. Price for her wonderful photographic record of current doings and diggings.

I am very grateful to Sam Weller, a long-time friend and valued advisor, who made the publishing of this third edition possible.